IMAGES
of America

PEABODY FIREFIGHTING

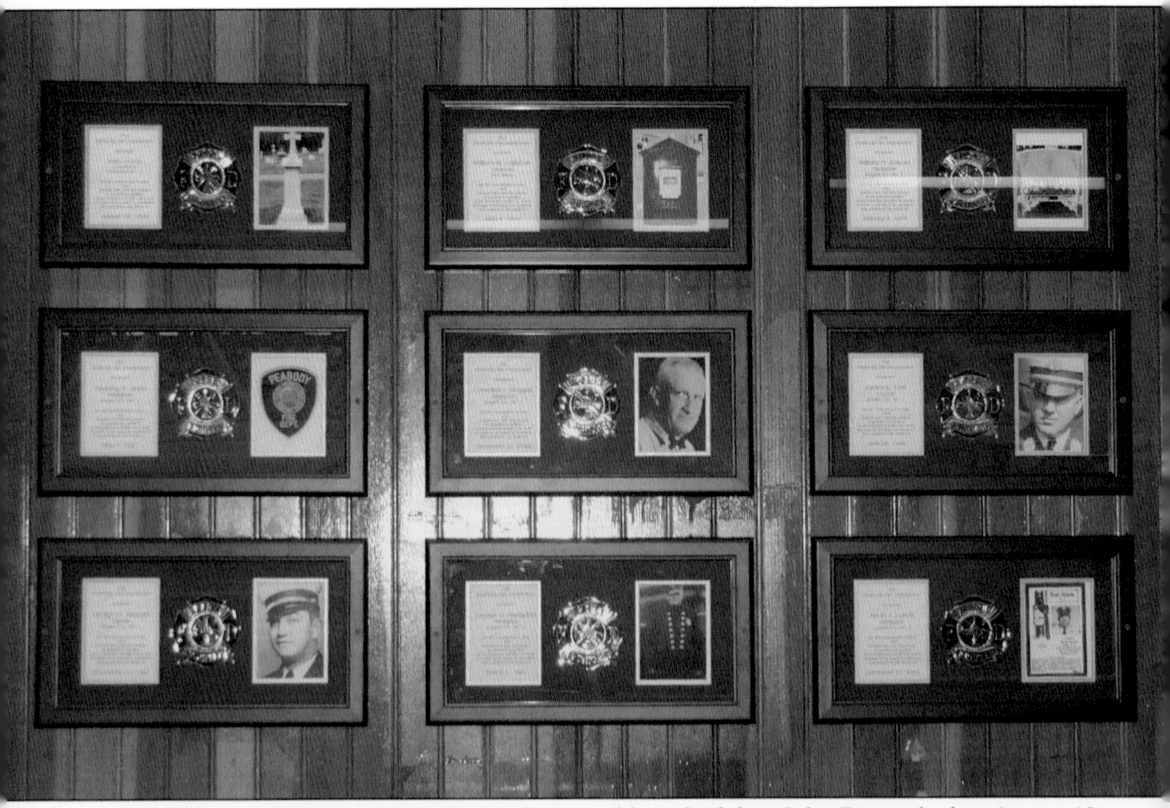

The following firefighters were killed in the line of duty: firefighter John Doran died on August 18, 1876; lineman William M. Callahan died on May 4, 1910; firefighter William H. Kerwin died on January 9, 1916; firefighter Timothy F. Shea died on May 7, 1931; firefighter Cornelius J. Duggan died on December 21, 1933; fire captain James Carr died on April 24, 1936; fire captain George H. Jacques died on November 14, 1942; firefighter George G. Hodgdon died on March 1, 1944; and firefighter Kevin J. Lynch died on December 10, 2001. This wall of honor is on display on the apparatus floor of Central Station.

On the cover: Please see page 59. (Courtesy of Peabody Historical Society.)

IMAGES
of America

PEABODY FIREFIGHTING

Ted Quinn

ARCADIA
PUBLISHING

ISBN 978-0-7385-5474-7

Published by Arcadia Publishing
Charleston SC, Chicago IL, Portsmouth NH, San Francisco CA

Printed in the United States of America

Library of Congress Catalog Card Number: 2007925451

For all general information contact Arcadia Publishing at:
Telephone 843-853-2070
Fax 843-853-0044
E-mail sales@arcadiapublishing.com
For customer service and orders:
Toll-Free 1-888-313-2665

Visit us on the Internet at www.arcadiapublishing.com

With special thoughts and in grateful memory of those who paid the ultimate sacrifice while in the performance of their duty, this book is dedicated to the men and women of the Peabody Fire Department, past and present, who have served their community with courage and devotion.

CONTENTS

ACKNOWLEDGMENTS

A special thanks goes out to my loving and supportive wife, Jill, my three beautiful daughters, Kayla, Anna, and Lilah, to my mother and father, Ed and Niki, my brother, Brian, sisters, Karen and Andrea, and to all those who helped, inspired, and put up with me.

I would also like to express my gratitude to the following individuals and groups that shared with me and granted me access to so many magnificent images, educated me with their knowledge, and inspired me to foster my tenacious nature: Deputy Chief James T. Coughlin; Jimmy Kimber; Billy Jacavanco; John (JDS) Gahagan; Ralph "Putty" Putnam; William C. "Bones" Mahoney III; George "King Snake" Nelson and John Nelson; Dave Cronin; Don Perry; John Wells; Nancy Barthelemy, Peabody Institute Library's archivist; Martha Holden, Peabody Institute Library's director; Peabody Institute Library staff and trustees; Chief Nicholas Gerakaris (retired); Chief Joseph Mendonca (retired); Chief Steve Pasdon; the chief's staff; the fire prevention staff; Barbara DeBruyckere; Donna Zellen; Rachael Walfield; fire alarm superintendent Mike Eliuk; Joe DaSilva; Nelson Burke; fire prevention captain David Sampson; Inspector Joseph Vinchens DiFranco Jr.; the Peabody Historical Society and its helpful volunteers; historical society president Bill Power; Anne Birkner; Heather Lovell; Barbara Doucette; Dan Doucette; George Peabody House Museum curator Marieke Van Damme; Rosemary O'Brien; Deputy John DeAngelo (retired); Uncle Bob Quinn; Neil Magesky; Charles Lawrence; Gloria Lawrence; Peter Fallon; Tom Martinuk; Russell Lewis; Roger Baker of Rehab 5; Tom Tremblay; the Peabody city clerk's office staff; city clerk Natalie Maga (retired); city clerk Tim Spanos; Colleen Kolodziej; Lisa Kent; Irene Zielski; Peabody Institute Library of Danvers archivist Richard Trask; freelance photographer Jim Daly; Marty Dullea; Henry Collins of the *Lynn Item*; Erin Stone at Arcadia Publishing; and everyone else that helped me make this albeit arduous but gratifying project a reality. If I left anyone out, please know that it was in no way intentional.

INTRODUCTION

Leather has always played an important function in firefighting. Fire buckets, fire hoses, and the harnesses and reins that were used on the horses that pulled the apparatus were all made of leather. The leather that was used by the Peabody Fire Department may have been produced in Peabody because the town was once nicknamed the "leather capital of the world." The town's location was ideal for leather tanning because of its many bodies of water, which were needed for tanning leather. This same water was also used to extinguish many tannery fires. Peabody, at the pinnacle of its status as king of the leather industry, had hundreds of leather factories and tanning shops that employed thousands built along its brooks, rivers, and downtown area. Today firemen still use leather gloves and leather boots, and some are even lucky enough to be owners of a leather helmet.

The leather industry in Peabody started to wane in the 1960s and hung on until the 1980s. The leather shops were forced out of business or forced to move to foreign soil because of increasing federal environmental mandates concerning the tanning industry as well as cheap labor that was found in third world countries. The mass exodus of the leather companies left many vacant and highly vulnerable factories saturated with centuries of volatile chemicals. These factories, made of large timber and with large open areas, were prime targets for arsonists and conflagrations-in-waiting. They were the subjects of many general alarm fires in Peabody.

Fire has plagued man since the dawn of time, and in Peabody it also showed no mercy. The kind of men who fought these fires and their unsurpassed inspiring courage would not change, but the way in which they fought the fires would; however, the basic principle is still the same, "putting the wet stuff on the red stuff."

Bucket brigades evolved into hand-pumped fire engine companies, and later on, with increasing innovations, the hose companies used high-pressure hydrant lines and steam boiler fire engines. Then came the introduction of the machine that would eventually replace the horse and expedite the arrival of the firefighters to the scene of the catastrophe: the internal combustion engine.

Everyone remembers the biggest, most spectacular fires. It is surprising how many people, even firefighters, have never heard of some of the more tragic fires. In Peabody, some of the biggest fires that caused the most damage, where one would expect the most casualties and fatalities, were often just "surround and drowns," as firefighters put it. It is the small house fire with the family of five that perished, the farmhouse on the outskirts of town where nine family members died while sleeping, or the schoolhouse fire where 21 young and innocent girls lost their lives in one horrific moment that truly left a scar on the citizens and firefighters but faded with the passage of time.

Firefighters, like most other people, deal with their emotions in many different ways. To let off a little steam, enjoy some camaraderie, or just have a little fun, firefighters attend yearly outings, the annual Firemen's Ball, and several dances and parties. These gatherings prove to have the desired effect.

Sit back and allow *Peabody Firefighting* to show the many faces and places of Peabody and its fire department. Included are images that capture moments in time that best depict how and where the firefighters confronted, fought, and inevitably beat down the fires. This book will show rare images of the buckets they used, the equipment they used, and the fires they fought. The names of firefighters pictured herein and their roles in the history of the proud occupation of firefighting will also be shared.

One

Fire Buckets and Hand Engines

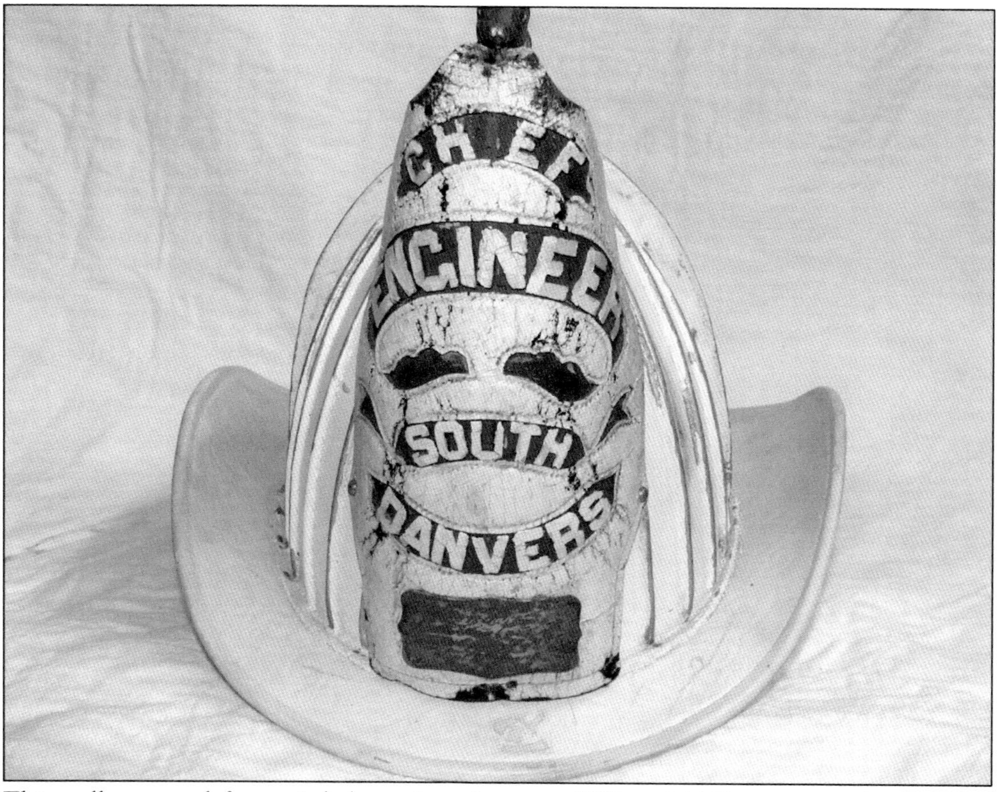

This well-preserved fireman's helmet, from between 1855 and 1868, was worn by the chief engineer of the fire department. It was made of leather, quite possibly from leather that was tanned in the leather city itself, Peabody.

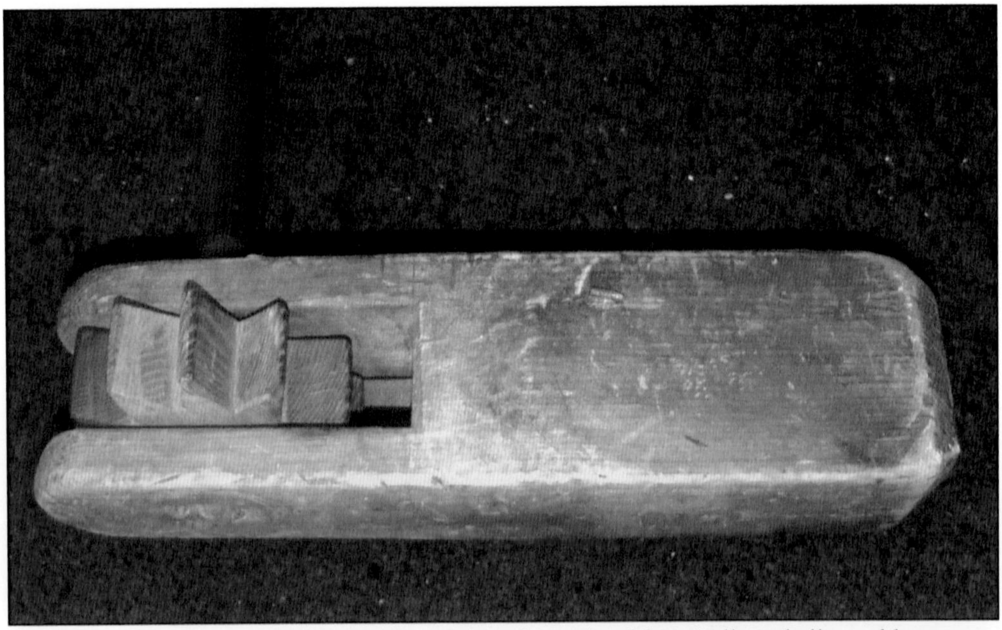

Noise, in one form or another, has always alerted people of fires. Yelling, bells, and horns, are often remembered, but this device has almost been forgotten. This wooden clacker sounded by a watchman, or town crier, was used to alert people of fire. It was held in one hand and swung in a circular motion, which caused the cogs to strike a wooden strip that made a sound similar to the plastic New Year's Eve party favors of today.

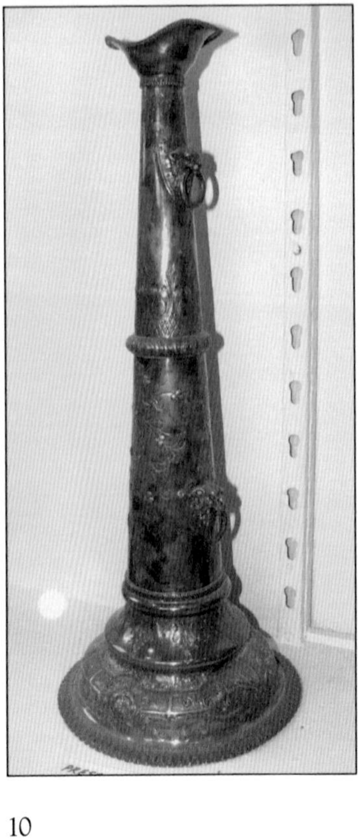

The trumpets, now only used to display the rank of an officer on his or her uniform, were once used to amplify the orders of the officer at the fire. This trumpet was a trophy won in a tank-filling contest by the Eagle No. 5 hand engine and crew at a firemen's muster in Lynnfield on July 18, 1860. The Eagle, purchased in 1836 for $575, was built by Edward Leslie of Newburyport. It was the fourth hand engine purchased by the town, and it was housed near the Eagle Pole at Main and Washington Streets, where it received the name Eagle.

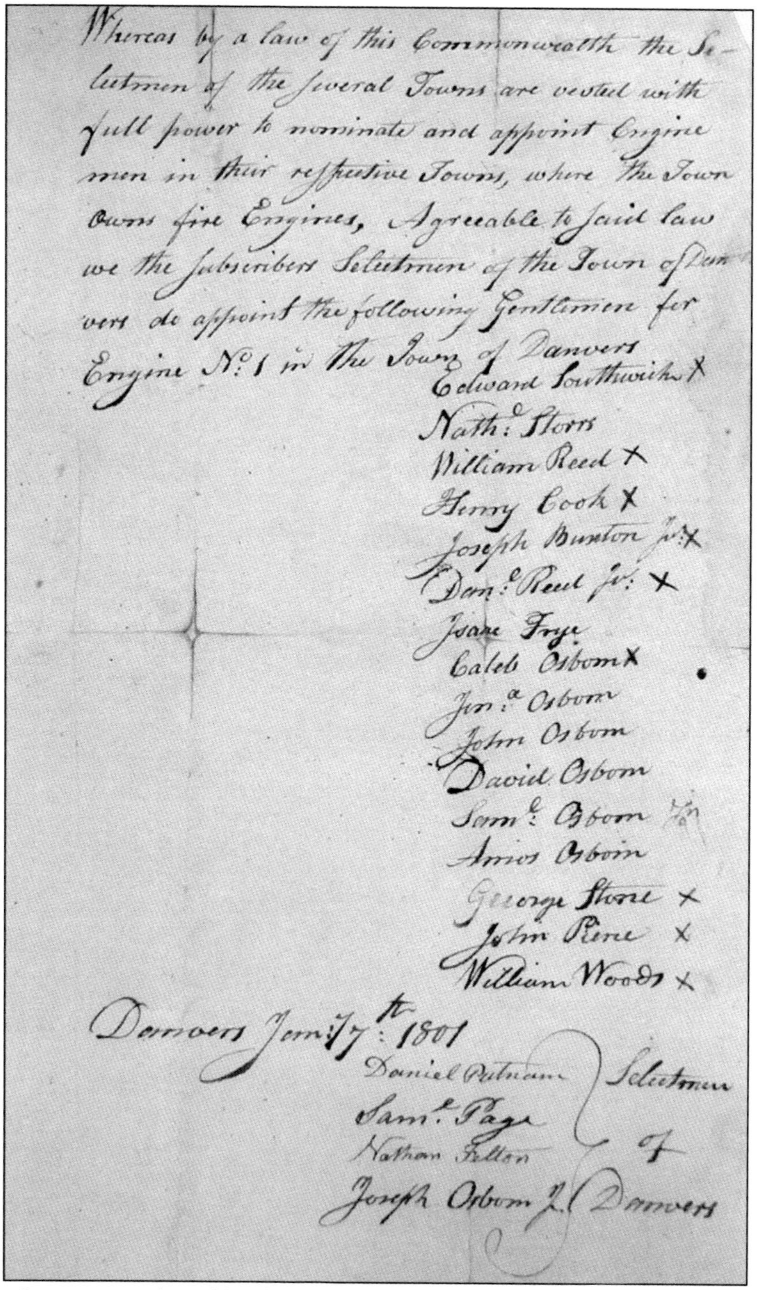

Whereas by a law of this Commonwealth the Selectmen of the several Towns are vested with full power to nominate and appoint Engine men in their respective Towns, where the Town owns fire Engines, Agreeable to said law we the subscribers Selectmen of the Town of Danvers do appoint the following Gentlemen for Engine No. 1 in the Town of Danvers

Edward Southwick X
Nath.l Storrs
William Reed X
Henry Cook X
Joseph Burton Jr. X
Dan.l Reed Jr. X
Isaac Frye
Caleb Osborn X
Jon.a Osborn
John Osborn
David Osborn
Sam.l Osborn
Amos Osborn
George Stone X
John Pierce X
William Woods X

Danvers Jan. 17th 1801

Daniel Putnam
Sam.l Page
Nathan Felton
Joseph Osborn Jr.
Selectmen of Danvers

The first hand engine purchased by the Town of South Danvers (now Peabody) was the Federal No. 1. It was a hand-pumping engine made by Ephraim and Stephen Thayer of Boston. Buckets were poured into its reservoir, and with its five-inch piston and men working the brakes, water would spray out of the hose onto the fire. It cost $500 and it served the town very well for many years. Federal No. 1 fought many fires during its service, including one of the biggest at the Crowninshield Woolen Mills on November 25, 1817. The Niagara No. 1 hand engine, built by the William Hunneman Company, replaced the old Federal No. 1 in 1822. This is the original roster of the first documented firemen of South Danvers. (Courtesy of Peabody Institute Library of Danvers.)

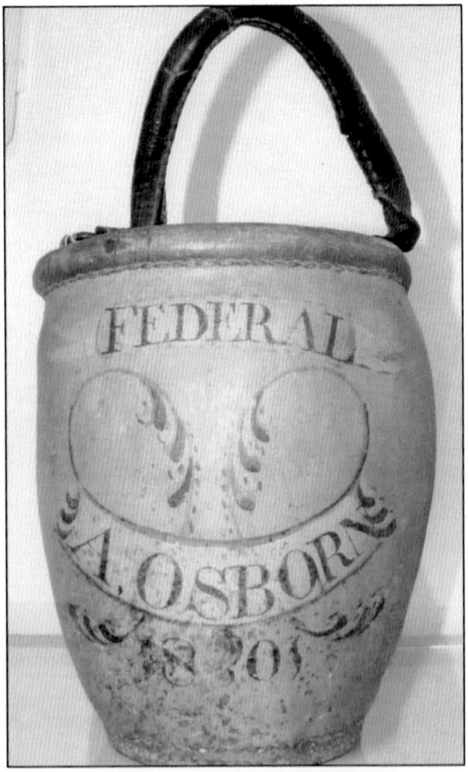

These fire buckets, now property of the Peabody Historical Society, are the actual fire buckets used by firemen in the 18th and 19th centuries. The bucket seen on the left belonged to Amos Osborne and is dated 1801. Osborne was a member of the Federal No. 1 engine company, the first hand engine in Peabody. The bucket seen below is from the Torrent No. 3, the third hand engine purchased by the town.

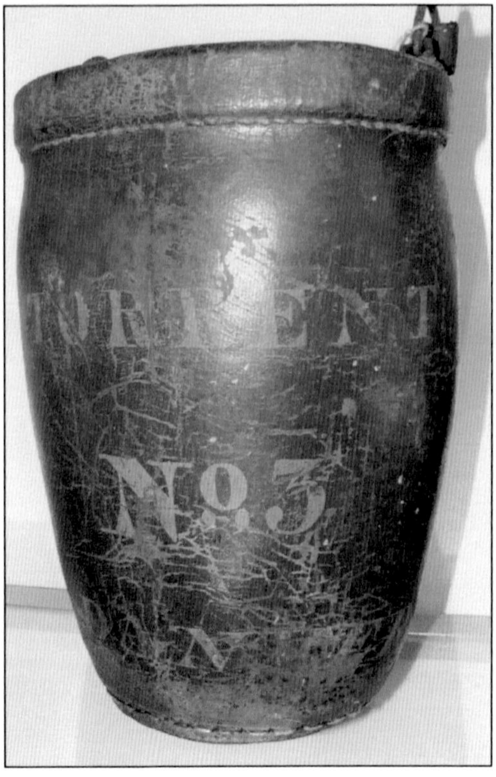

Peabody, like other cities, developed a water system using bored-out crude logs that were tapered and fitted together. Every half block or so, a wooden plug was inserted into this water system, and when water was needed, this plug was removed. The term *fireplug* was derived from this system. In Peabody, as recently as the late 20th century, some of these wooden logs were dug up when replacing water lines. This one was found under Main Street.

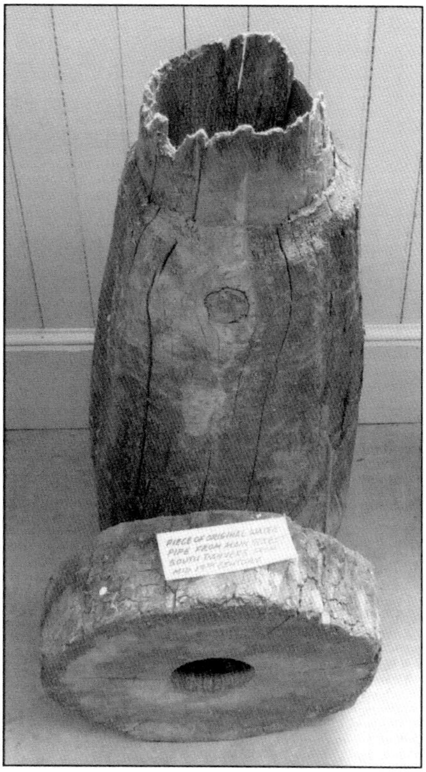

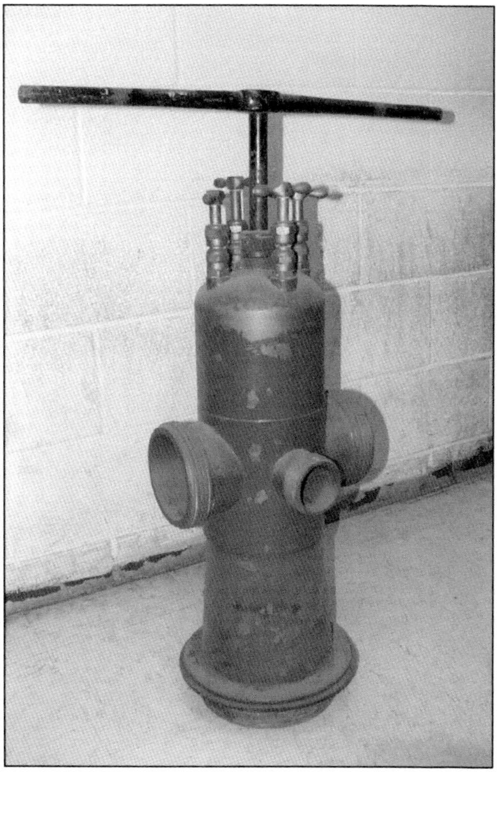

Hydrants come in different colors, shapes, and sizes for many different reasons. A Lowery hydrant was a portable hydrant that was either carried on a fire truck or placed in a convenient location. It was screwed into a fitting that was flush or somewhat recessed in the ground, and when a valve was turned on, it worked like a normal hydrant. In 1908, a Lowery hydrant chuck, similar to this one from Salem, was used on a hydrant in Peabody Square and located in a small closet in the western end of the front steps of the South Church, where it was convenient to get to when needed. The problem with these hydrants was that if there was snow on the ground or a vehicle was stopped over it, the firemen would not know its exact location.

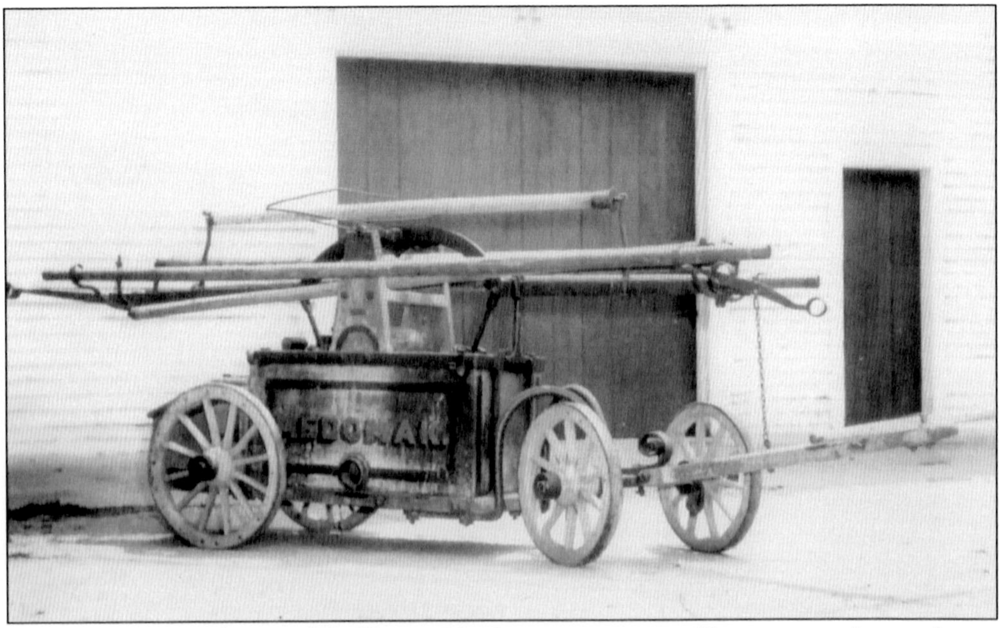

"Always ready" was the cry of the Torrent No. 3 engine company. *Torrent* means "a tumultuous flow of water," and that was the case for this engine. Its membership was filled with eager firemen and was commended on many occasions. It was the third engine that the town purchased and was built by the William Hunneman Company of Roxbury in 1830. It had a six-inch pump and cost the town $575. Its first location was at the railroad freight yard on Central Street near the square.

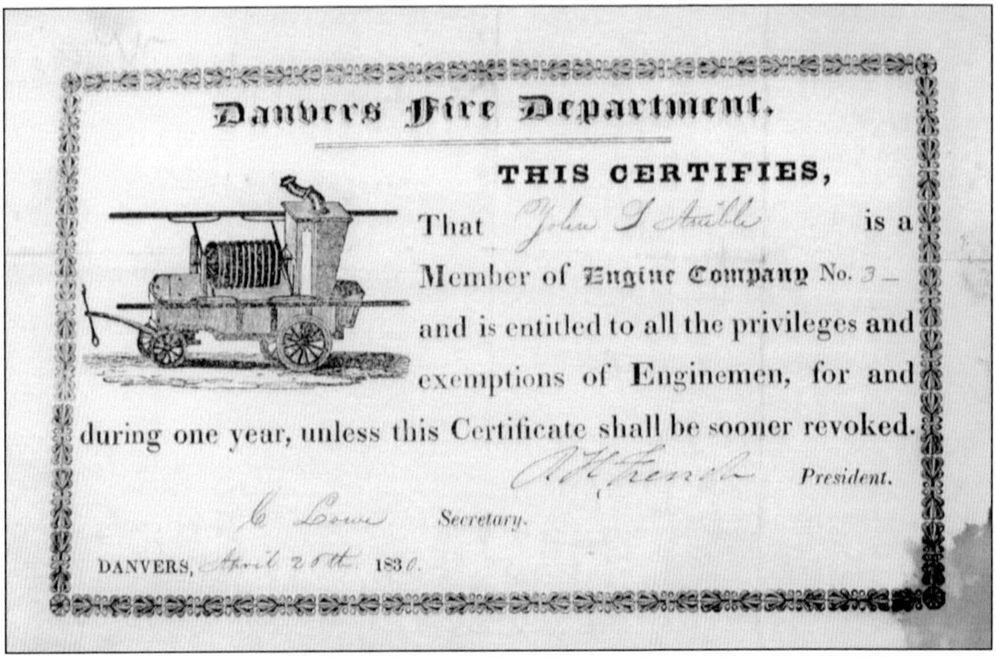

Danvers Fire Department.

THIS CERTIFIES,

That John S Anible is a Member of Engine Company No. 3 and is entitled to all the privileges and exemptions of Enginemen, for and during one year, unless this Certificate shall be sooner revoked.

A H French President.

C Lowe Secretary.

DANVERS, April 20th 1830.

Certificates like this one were usually signed by the president and secretary of the engine company. This certificate is dated April 20, 1830, and certifies that John Anible is an engineman of Torrent No. 3 for one year.

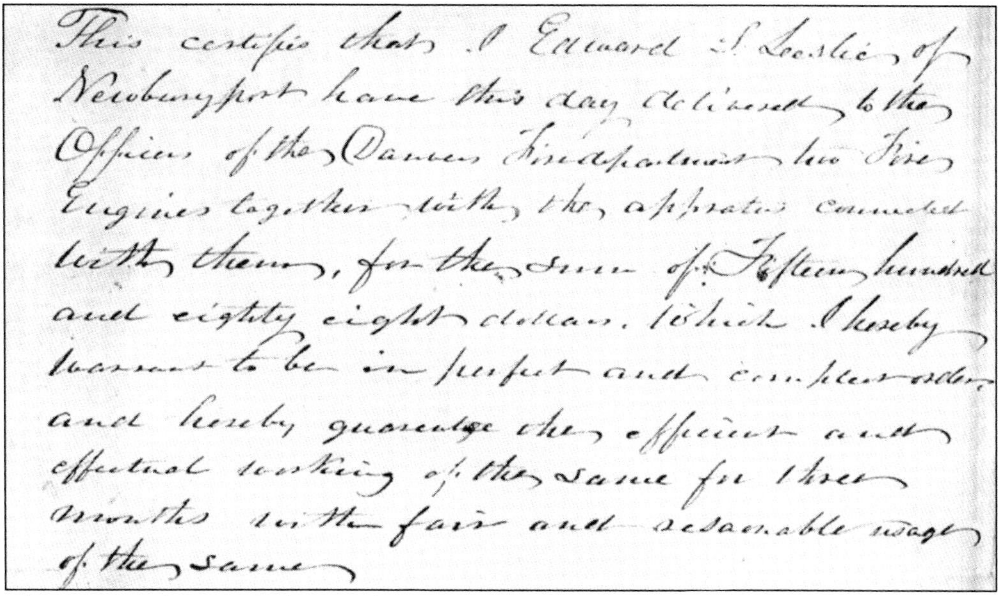

This certifies that I Edward S. Leslie of Newburyport have this day delivered to the Officers of the Danvers Fire department two Fire Engines together with the apparatus connected with them, for the sum of Fifteen hundred and eighty eight dollars, which I hereby warrant to be in perfect and complete order and hereby guarantee the efficient and effectual working of the same for three months with fair and reasonable usage of the same

The next "masheen," which was what the old firemen called their hand engines, was built by Edward Leslie of Newburyport and named the General Foster No. 7. It was purchased in 1843, just after the great conflagration of South Danvers in the area of what is now Peabody Square, and was delivered on March 1, 1844. The engine was named after the Peabody native and Revolutionary War general Gideon Foster. It was later renumbered No. 2 after Peabody separated from the town of Danvers. This is a copy of the receipt from Edward Leslie.

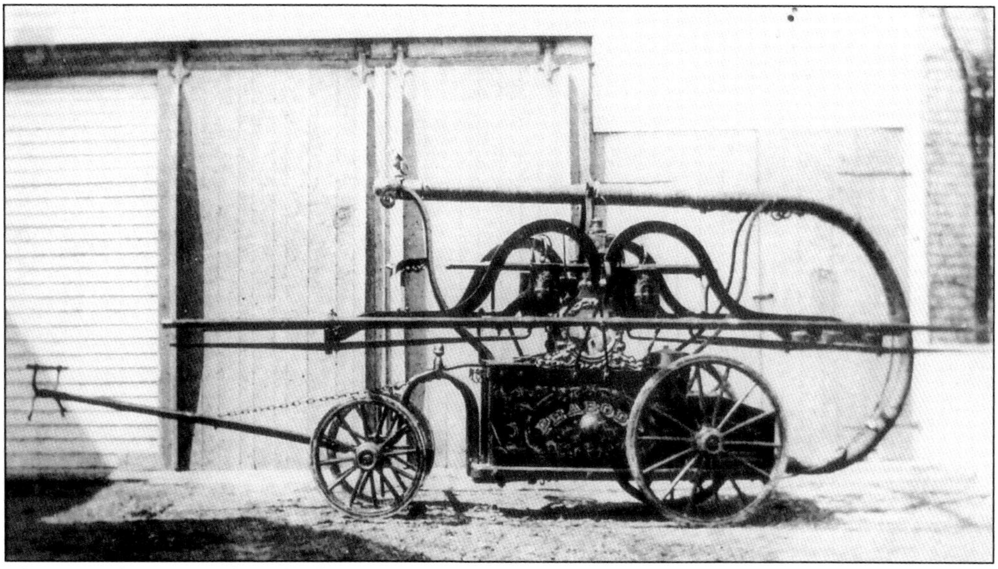

The second engine to be called the Torrent No. 3, replacing the first one, was also an engine built by the William Hunneman Company, and the manufacturer's serial number was No. 363. It was purchased in 1849 at a cost of $694. It was a side-stroke engine with a six-and-one-eighth-inch cylinder. It retired on May 1, 1883, and was sold to Turner Falls. It returned to Peabody in 1889 and was renamed the Peabody by the Peabody Veterans Firemen Association, which was one of the first muster groups formed. In 45 musters, the engine won 21 prizes totaling $1,900. It was finally scrapped in 1910.

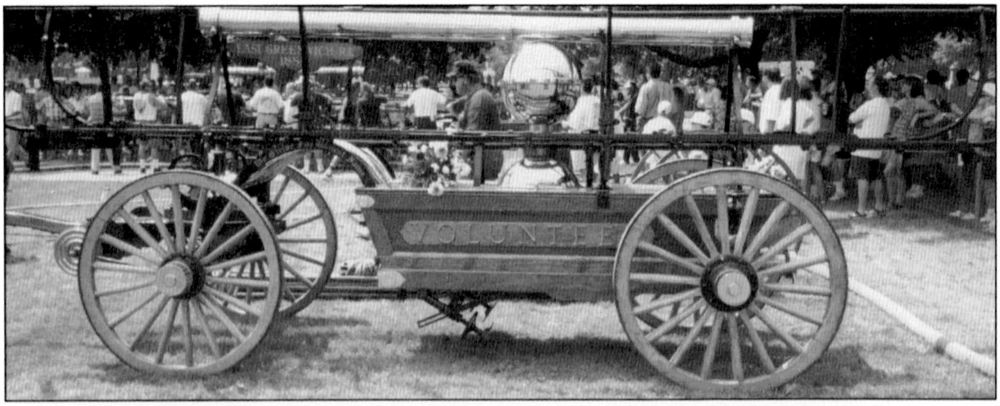

The first of two engines named the Volunteer No. 4 was built by the L. Button Company of Waterford, New York, in 1845. It was donated to the town by prominent businessman and army general William Sutton. It was housed on Pierpont Street near the general's house and served that part of town well. Its proximity to Salem made it possible to assist the town in times of need. A new volunteer engine was built and purchased in 1854, and the first one was returned to Sutton. This is a picture of the second engine, which is now cared for by the East Greenwich Firemen's Association in Rhode Island.

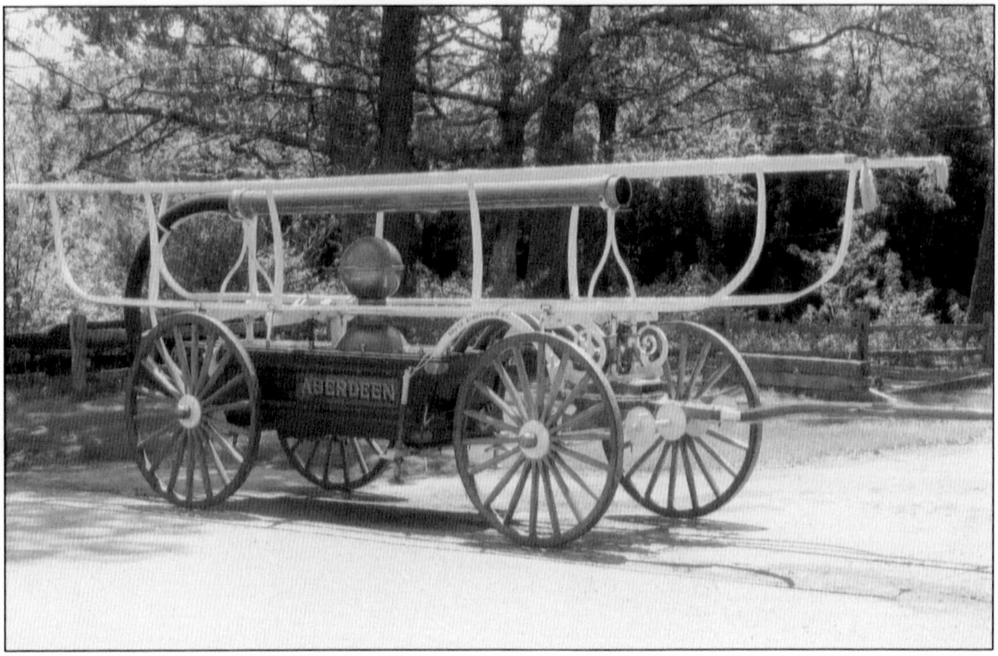

The last hand engine to be purchased by the town was the Sidney C. Bancroft No. 5, named after Sidney C. Bancroft, who was instrumental in getting an engine house built in the southern part of town, known as Rockville, which today is South Peabody. The engine was built and purchased in 1876 at a cost of $1,620. It was housed in its newly built quarters on Lynn Street. This station stayed in service until 1961 when it burned to the ground. A new station was built in its place. The engine was sold many times before finally settling in 1965 in South Kingston, Rhode Island, where it was renamed the Aberdeen, which was the name of the town's old tub. In 1966, it was the New England League Class A champion. It is still located and wonderfully cared for in South Kingston.

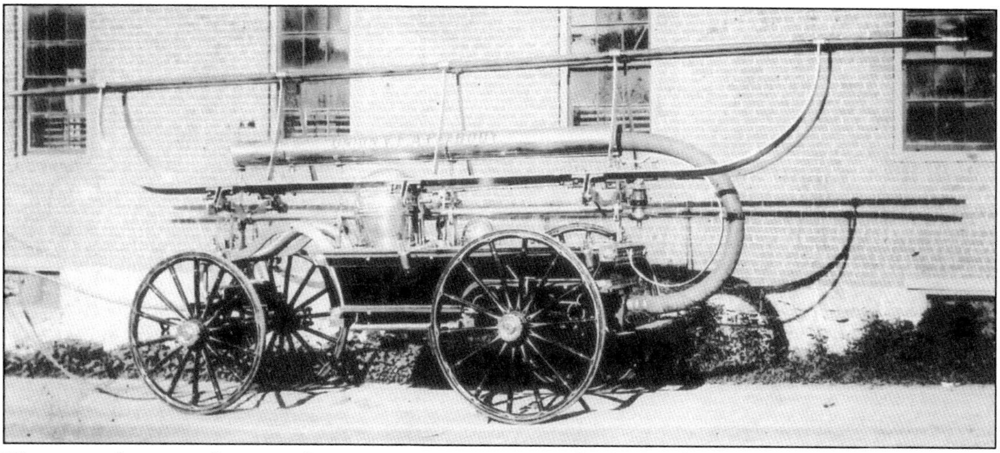

The second group of men to form an association in Peabody, for the purpose of firemen musters, was the Union Veteran's Association (UVA). The UVA was incorporated on February 16, 1893, and meetings were held every month at the engine house on Walnut Street. The association bought the Washington hand engine for $350 from Belfast, Maine. The engine was originally built in 1854 for Charlestown by William Jeffers of Pawtucket, Rhode Island. It cost $3,500. C. H. Hooper bought the hand engine from the UVA when it disbanded around 1906 and sold it to the King Philip Fire Engine Company of Bristol, Rhode Island. It was a record-breaking engine, with a long, credible history. It took part in over 50 musters and won 39 prizes totaling $4,917. Its record play was on October 20, 1898, in Nashua, New Hampshire, when it threw a stream 224 feet 2⅛ inches. It won first place many times, took many cash prizes, and once won a silk flag that was donated to the Peabody Historical Society when the engine disbanded. Its last muster was in Salem on September 11, 1906.

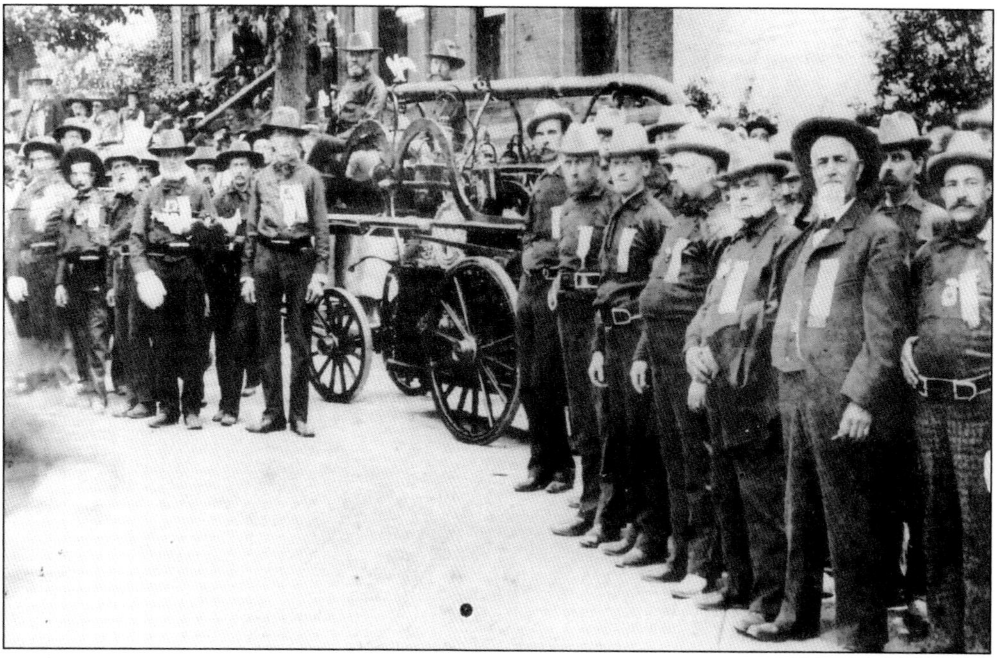

In this picture taken behind city hall, these men of Peabody's first muster group, Peabody Veterans Firemen Association, looking like cowboys, displayed with honor their hand tub and ribbons they had won at a fireman's muster.

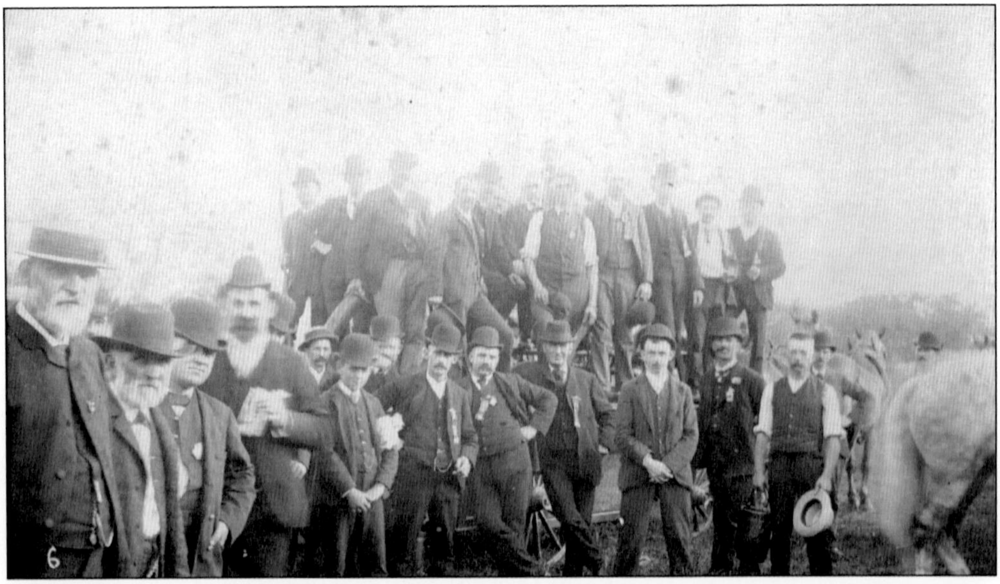

Rockdale Park in South Peabody was the location of a few firemen's musters in Peabody. Here the men pose for a moment with pride in themselves and their engine.

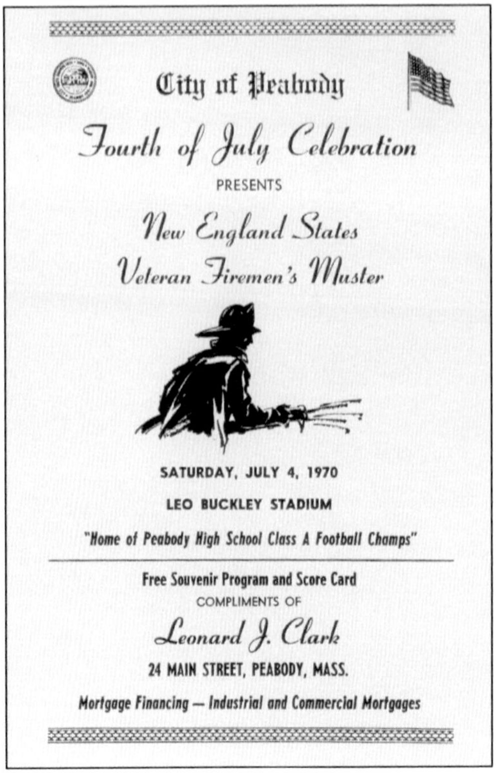

City of Peabody

Fourth of July Celebration

PRESENTS

New England States

Veteran Firemen's Muster

SATURDAY, JULY 4, 1970

LEO BUCKLEY STADIUM

"Home of Peabody High School Class A Football Champs"

Free Souvenir Program and Score Card

COMPLIMENTS OF

Leonard J. Clark

24 MAIN STREET, PEABODY, MASS.

Mortgage Financing — Industrial and Commercial Mortgages

On Saturday, July 4, 1970, the City of Peabody hosted the New England States Veteran Firemen's Muster. It was held at the Leo Buckley Stadium behind the high school on Central Street. This was the last firemen's muster to be held in Peabody. This is the cover page for the program of that muster. The Volunteer No. 4 took first prize in the first-class division that day and the Aberdeen placed seventh. On December 14, 1923, the bleachers of the stadium collapsed due to a fire underneath them. Many people were injured.

Two

The Engine Houses

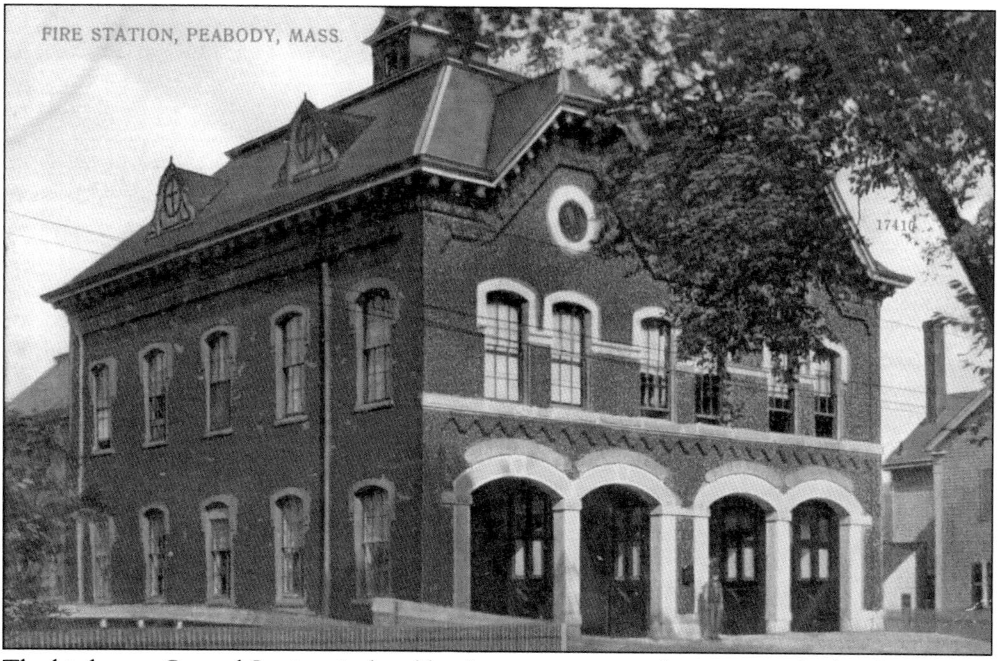

The big house, Central Station, is the oldest known continuingly operating fire headquarters in the country. It is listed as a national historic site.

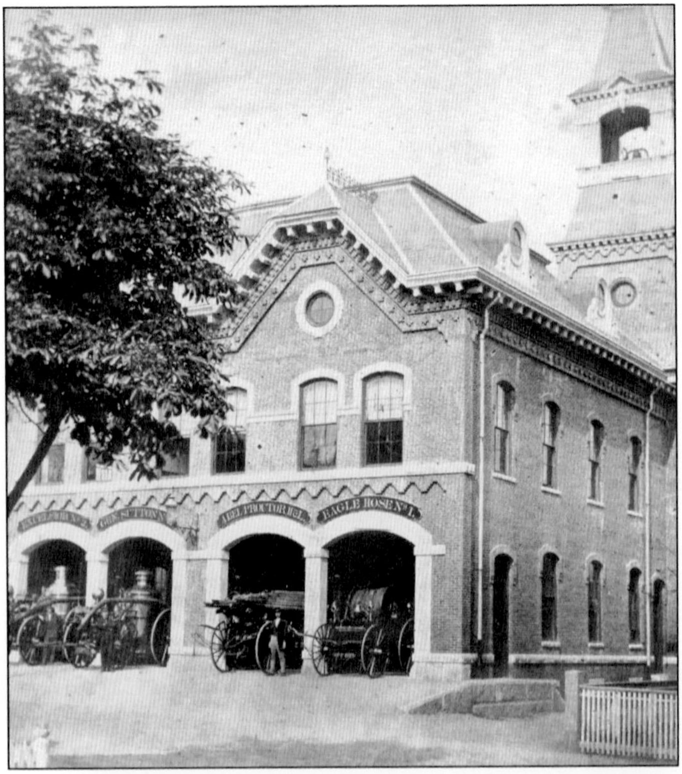

Built in 1873, Central Station housed the Abel Proctor No. 1 ladder truck, the General Sutton No. 1 steamer, Eagle Hose No. 1, and the Excelsior No. 2 steamer. Social events were held at each of the engine houses to raise money and to have some fun. In the days before the Firemen's Ball, Central Station held dances called the Three One and a Two.

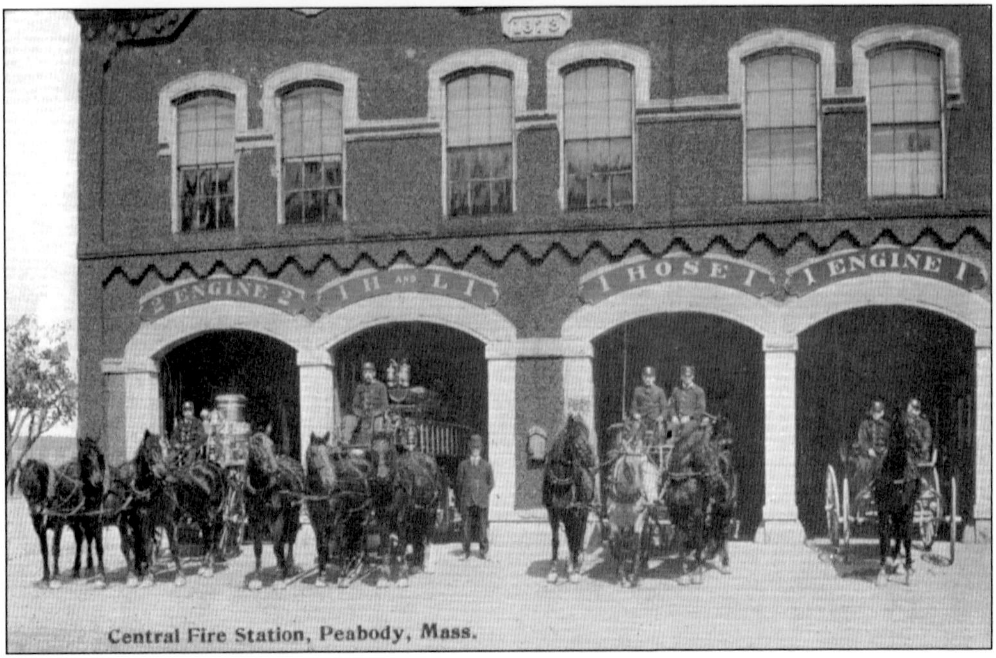

City scenes were popular subjects for real-photo postcards such as this one of Central Station. The generic signs over the doors replaced the original signs that named the apparatus, as the previous picture portrays.

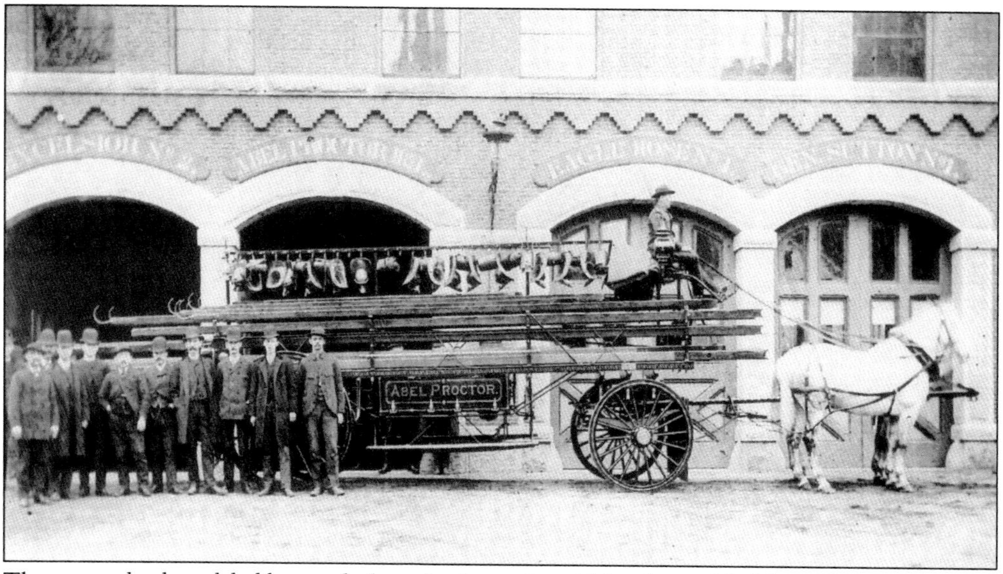

The name *hook and ladder truck* dates back to the old hooks and ladders of Colonial times and early days of the 19th century. The hooks were used to pull down buildings so the fire would not extend to other buildings, and ladders were used to access high places. Peabody's first horse-drawn hook and ladder truck was purchased in 1870. It was originally housed next to the General Sutton No. 1 steam engine at the corner of Foster and Church Streets. It was originally named the Miles Osborn and then renamed the Abel Proctor. Both Osborn and Proctor were well-respected firemen.

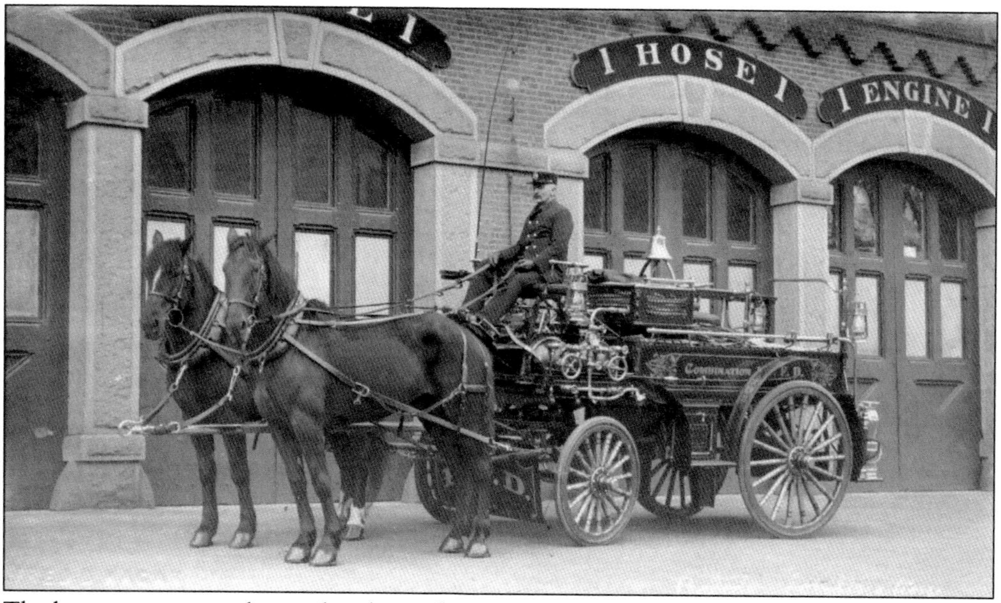

The hose wagons were heavy-duty horse-drawn wagons that supplied the firemen with hoses at the fire scene. The hose wagons worked in conjunction with the steamer engines, or the firemen would just attach the hoses to fire hydrants using high-pressure water to fight the fires. The hose wagon evolved into what was called a combination wagon, which not only supplied hoses but also had a chemical booster tank used to extinguish fires. Pictured here is combination No. 1 with its hose wagon and nickel-plated Babcock chemical tank under the driver's seat.

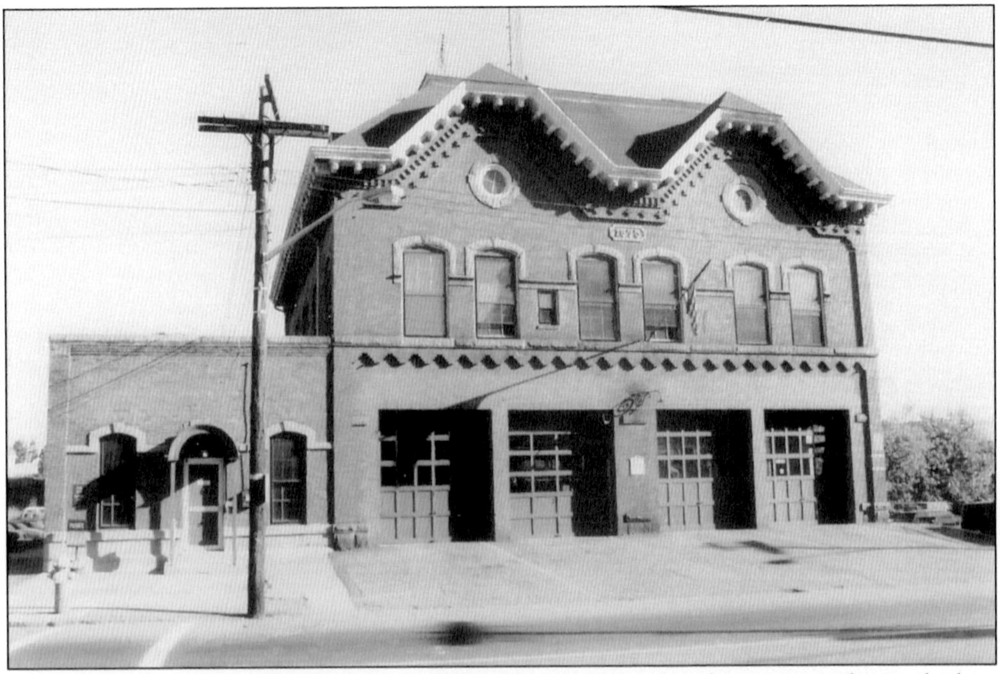

Central Station is pictured here prior to the historical and structural renovation that took place in 1988. Notice the squared-off doors that had replaced its beautiful granite arches.

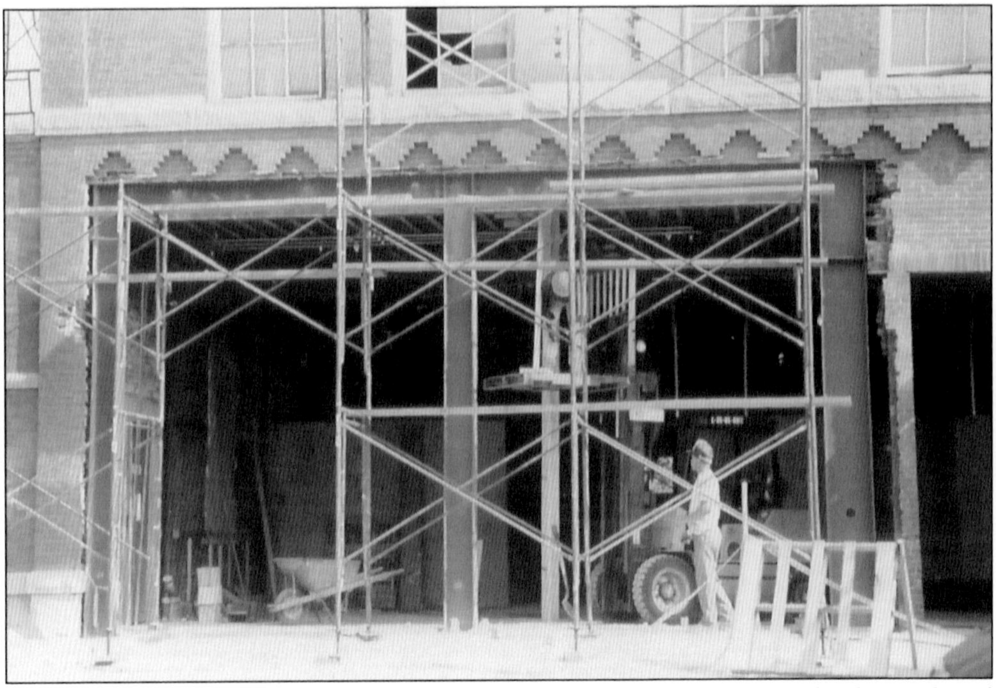

The remodeling of headquarters was no easy task. Most of the building needed to be reinforced with steel columns and headers. This picture shows the steelwork used to replace the old arched doorway.

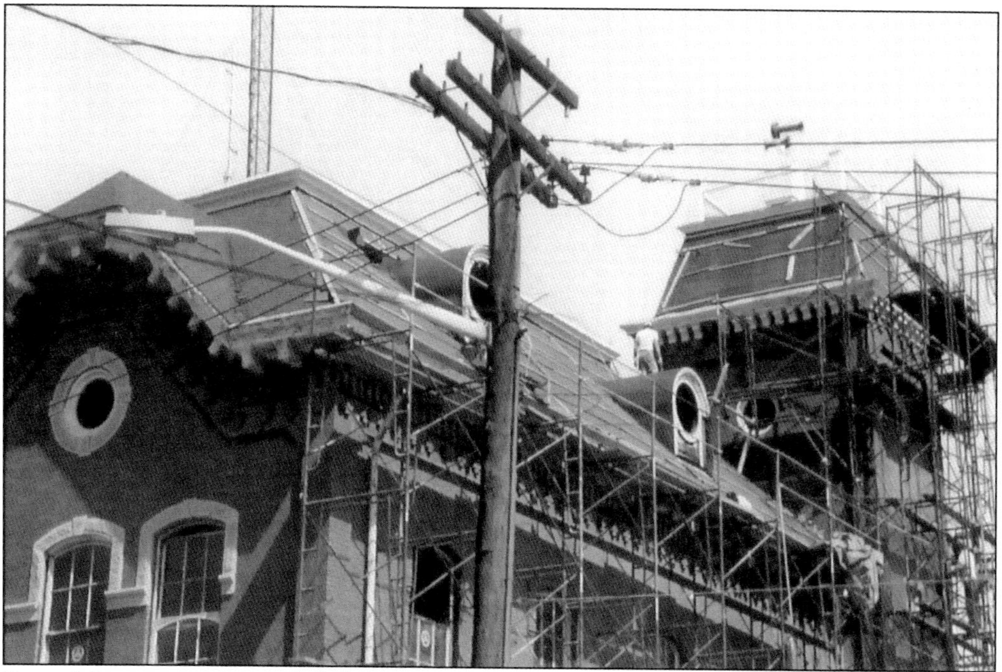

Part of the renovations included the repointing of the brick exterior and the reroofing of the entire building.

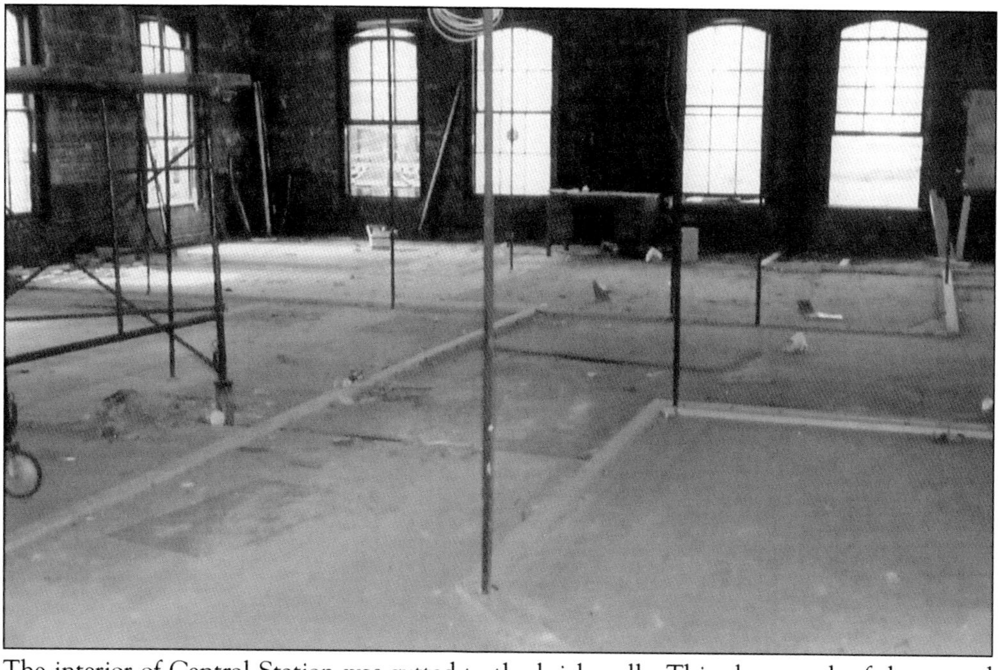

The interior of Central Station was gutted to the brick walls. This photograph of the second floor reveals steel rods that hung from the roof timbers through to the second floor to the first-floor ceiling. These rods held up the entire second floor and gave the first floor, where the apparatus was stored, four bays of unobstructed space. During the renovations, the steel rods were reattached to the steel reinforcement in the attic.

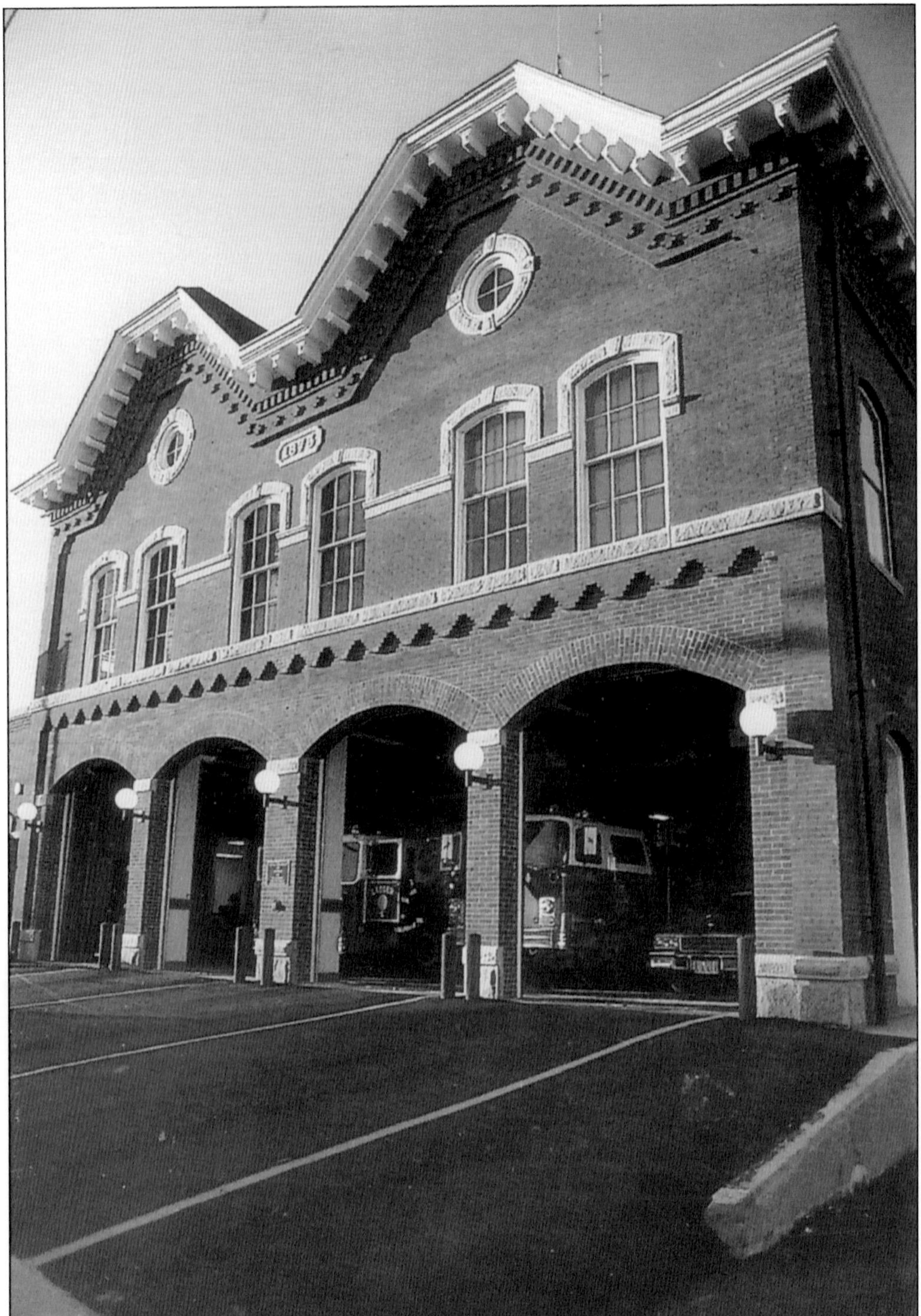

The finished restoration of Central Station in 1989 revealed the old arches, although not granite, and showcased a building bursting with the pride of its century-long history. The original cost of the station in 1873 was $30,000. The renovations in 1988 cost $1.5 million.

This artist rendition of the new Engine House No. 2, at the corner of Washington and Mason Streets, was in the *Salem Evening News* with an article about the opening of the new engine house and the dedication that took place on January 31, 1893. It was built in 1892 on the footprint of an old engine house that was located on the corner of Main and Wallis Streets and moved to this location in 1857. While on Main Street, the station housed the Eagle hand engine, and when it moved, it housed the General Foster No. 2. The station, like all firehouses, had a bell tower. Engine No. 2's bell tower had a weather vane with a fireman on it that may have been taken from the old station on Pierpont Street. This station was closed during the Depression and was razed in 1954. It is still a vacant lot.

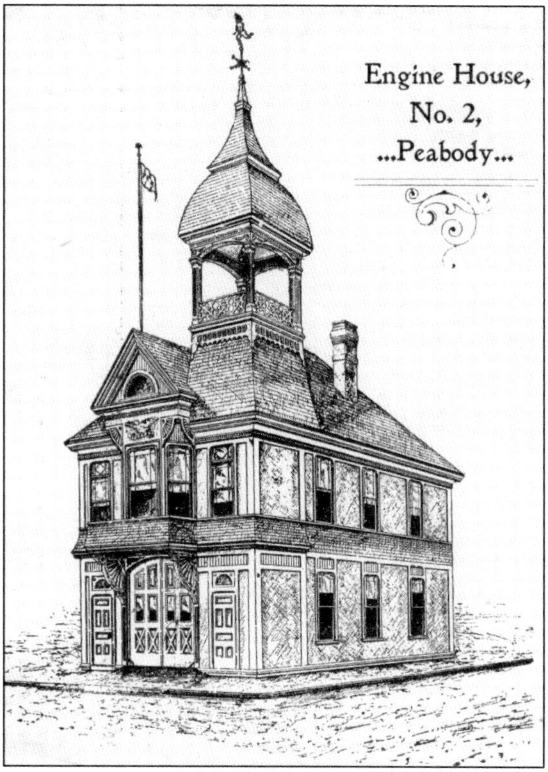

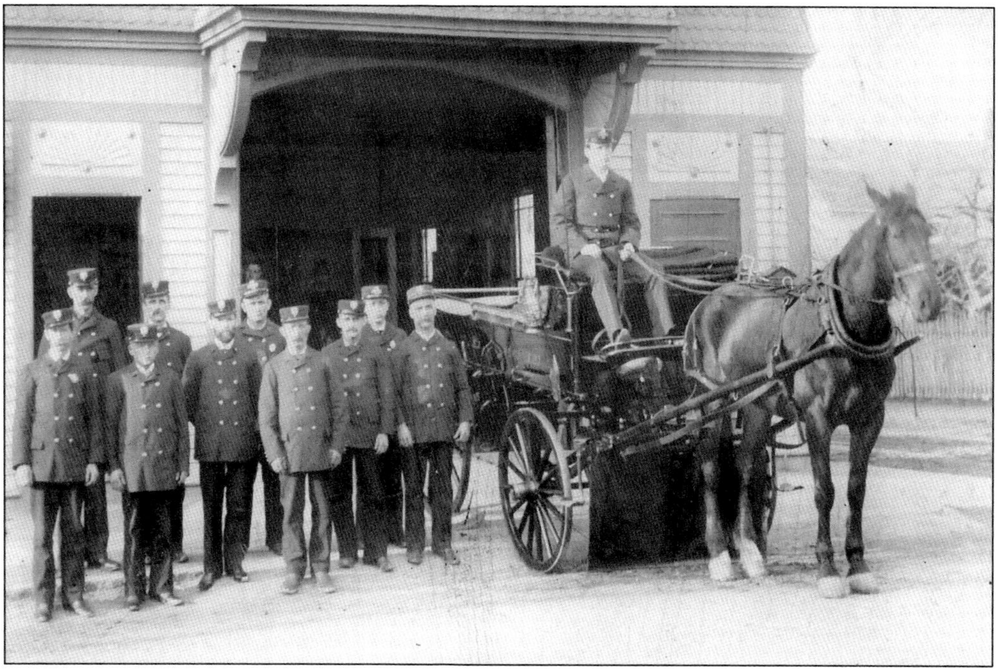

Shown here, from left to right, are firemen John Dunney, Calvin Larrabee, Patrick Ducey, Andrew Gray, H. G. Upton, John Calvert, Edward Boxwell, John Costello, Charles Berry, Frank Mason, and Lewis Buxton.

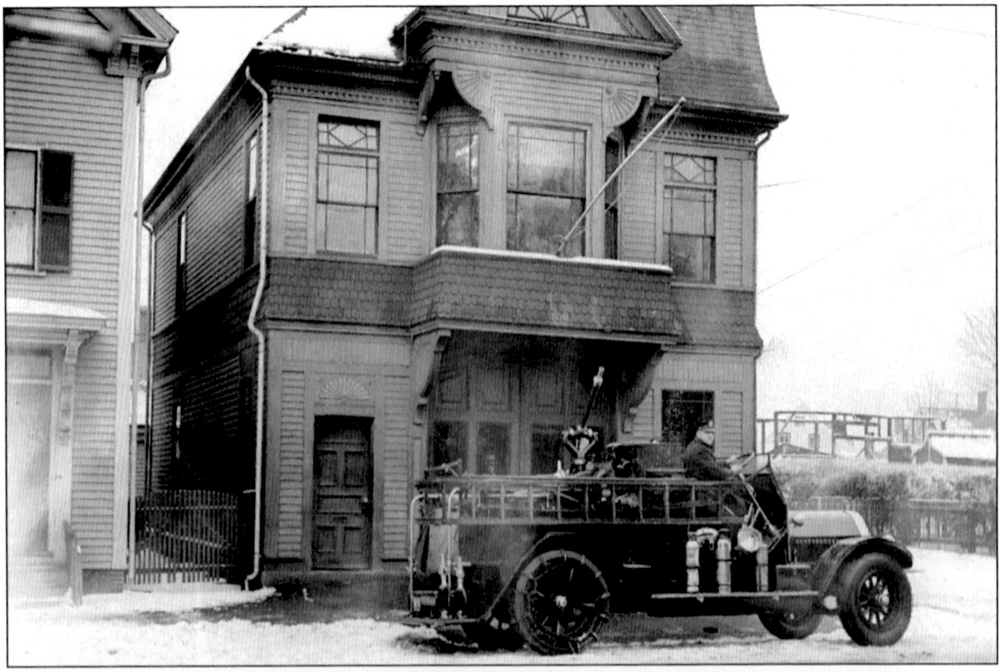

Engine No. 2., a Knox engine costing $5,000, arrived on July 21, 1915. It was bright red or scarlet and had "Combination 2, Peabody Fire Dept." painted on it. When it was tested, it drove around town with 17 men in the wagon.

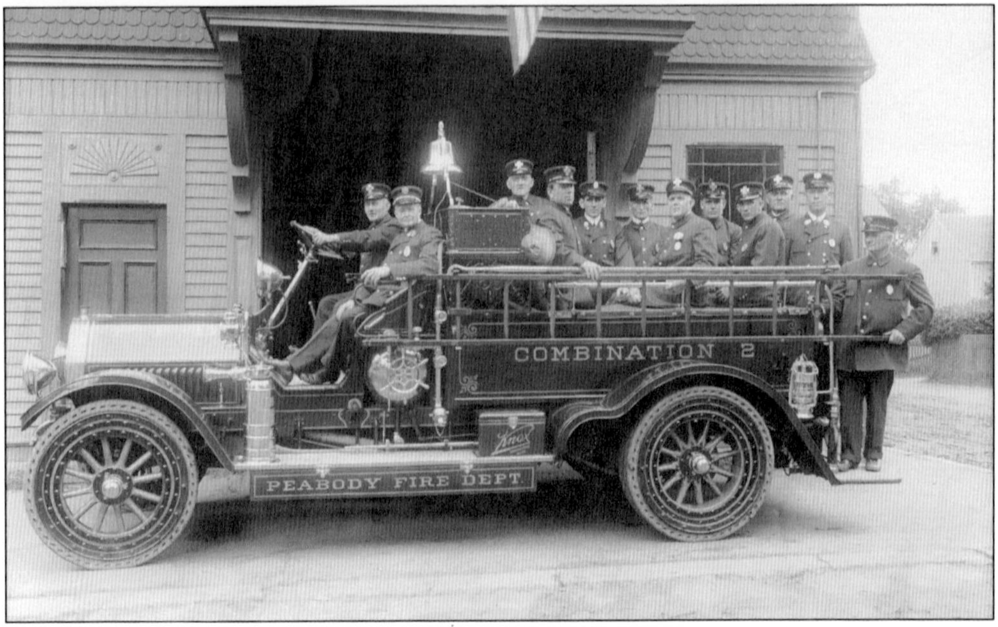

Engine No. 2 is seen here with engineer Thomas Hutchinson and his men. Engine House No. 2 was closed during the Great Depression and intermittently reopened. It was then used as a Works Progress Administration office and a voting location but was finally sold on December 10, 1953, for $275 and was razed in August 1954. Today the land where the station was located is still a vacant lot.

The Torrent No. 3 engine house moved five times between the years 1830 and 1876. It was first housed in 1830 at the old freight yard behind the hearse house on what is today called Railroad Avenue. When the second Torrent arrived in 1849, the engine house moved up Central Street to a new location. With the invention of the steamer engines and use of hose wagons, the Torrent engine house became the hose No. 3 engine house. It moved twice more before ending up on Endicott Street, pictured here with its hose carriage.

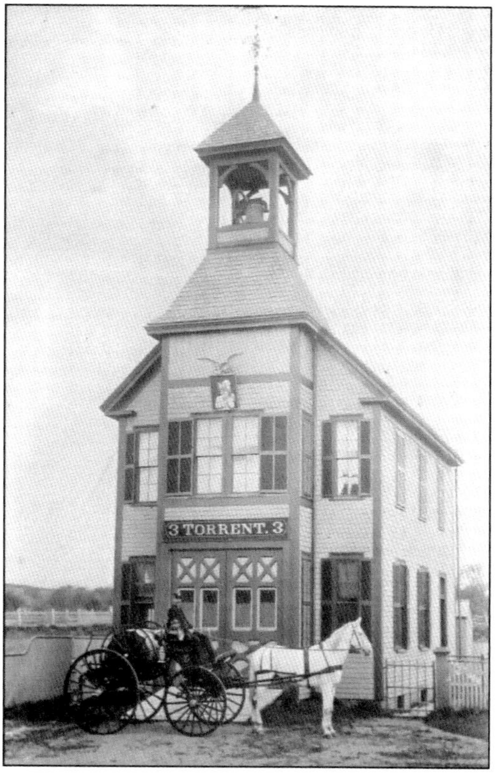

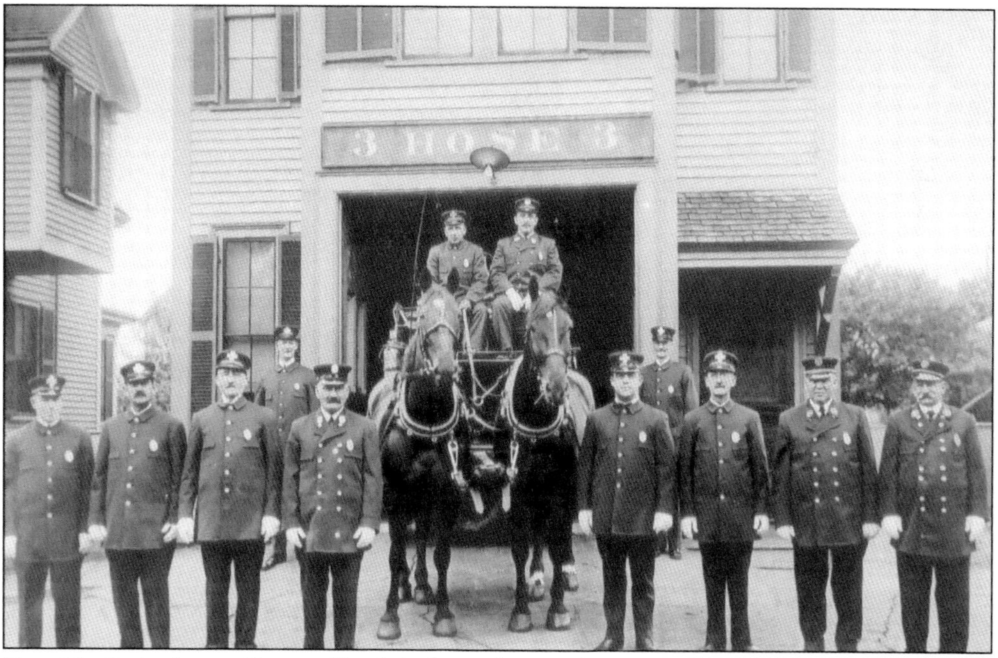

Hose No. 3, the former Torrent No. 3, is shown here with its men. From left to right are Fred W. Bodge, Gustave Bagley, Frank Toye, George Bodge (rear), Herbert Buxton, Freeman Bodge (seated at the reins), Benjamin Bodge (seated), horses Tom and Jerry, Lewis Saxton, Frank Durham (rear), Samuel Watts, Arthur Bodge, and Chief Jesse Barrett.

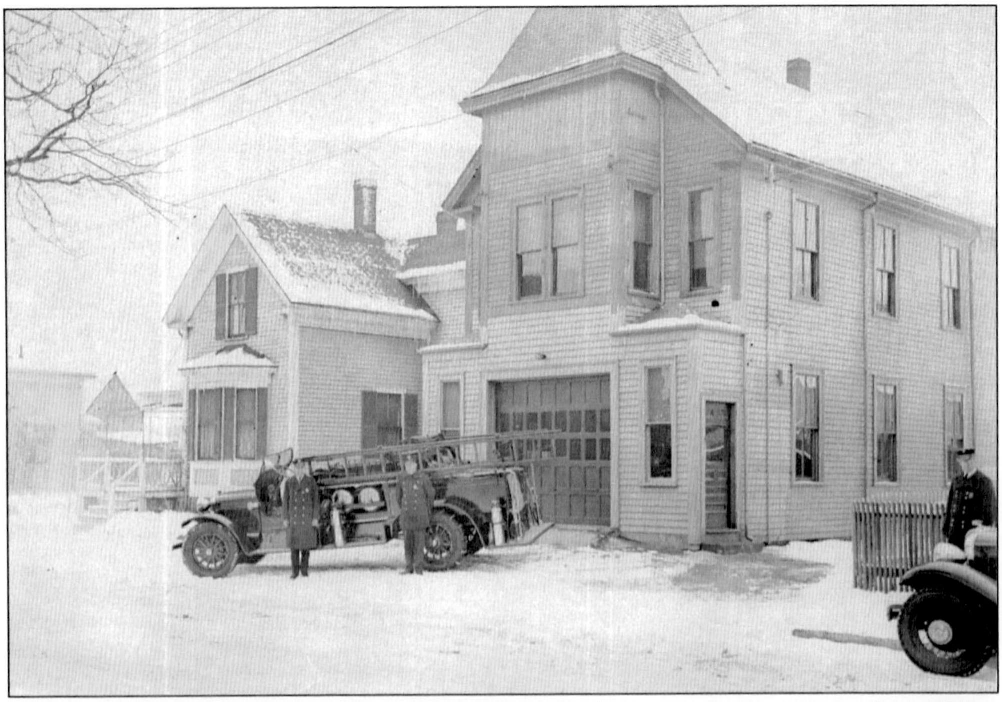

Here is a winter scene at Engine House No. 3, which was located at Endicott Street from 1875 to 1986. This engine house was closed in 1986, and the building was sold to the Peabody Historical Society for a museum. The location is now a vacant lot. A new station for engine No. 3 was built on Prospect Street near the North Shore Shopping Center.

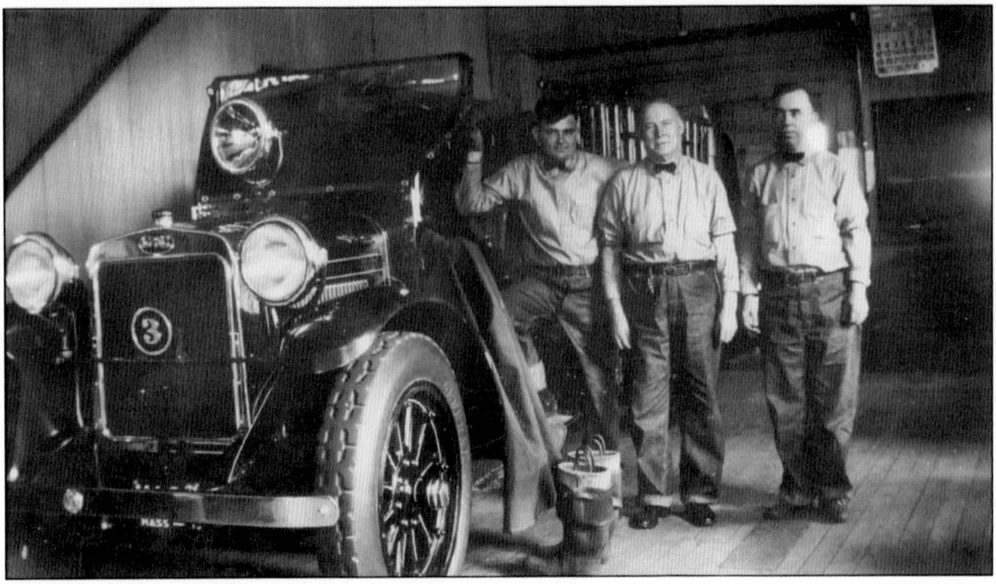

Engine No. 3, located on Endicott Street, is seen here with the crew. It was a 1926 GMC mounted on the old engine No. 1's chassis, a 1913 Seagrave. The 1913 Seagrave chassis was the first motorized engine that Peabody bought and it was used on engine No. 3, which was the last engine to use horses.

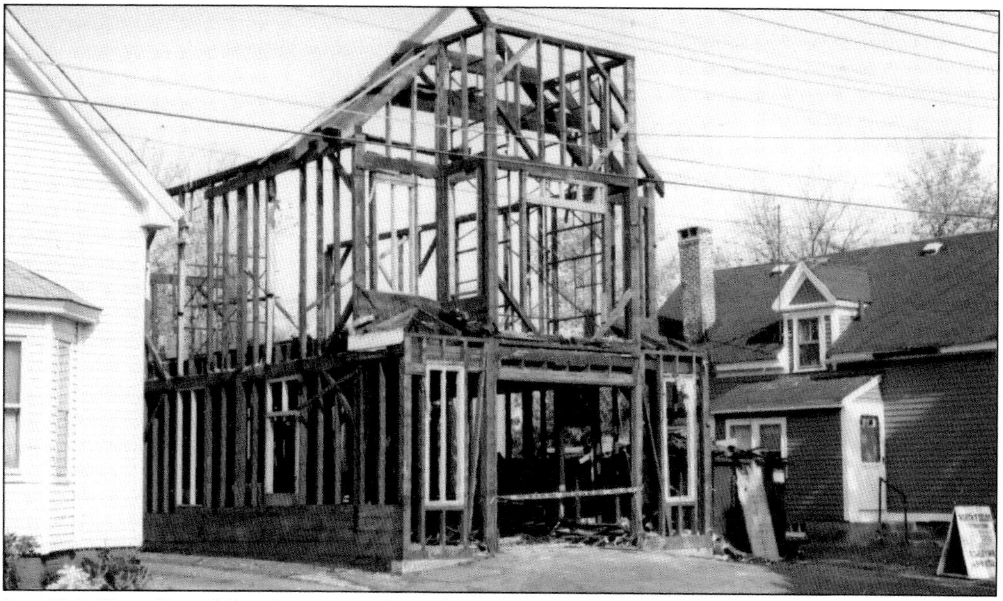

In 1988, Rosa Drysdale, the newly elected president of the Peabody Historical Society, sought to acquire the engine No. 3 engine house as her first act. In 1991, a committee of Drysdale, Joe Donlon, John DeAngelo, and Jim Stamatelos was set up, the city sold the building to the historical society, and the engine house was dismantled board by board. With the help of the society members and volunteers, construction began in 1993, and it was finally completed in 1996.

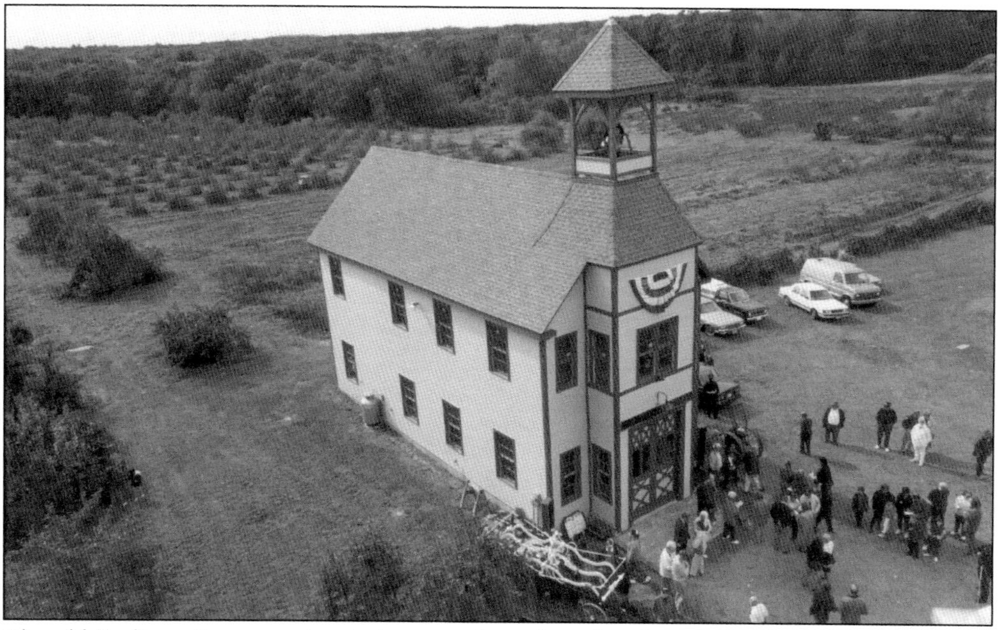

The old Endicott Street station, being torn down board by board, was reassembled like a big jigsaw puzzle at 38 Felton Street at Brooksby Farms. This first-class museum, owned and operated by the Peabody Historical Society, has many fine pictures, firefighting artifacts, and antiquities from Peabody. It houses a hand engine from Lynn, Tiger No. 1, and the second floor is a spacious hall that can be rented out for special occasions.

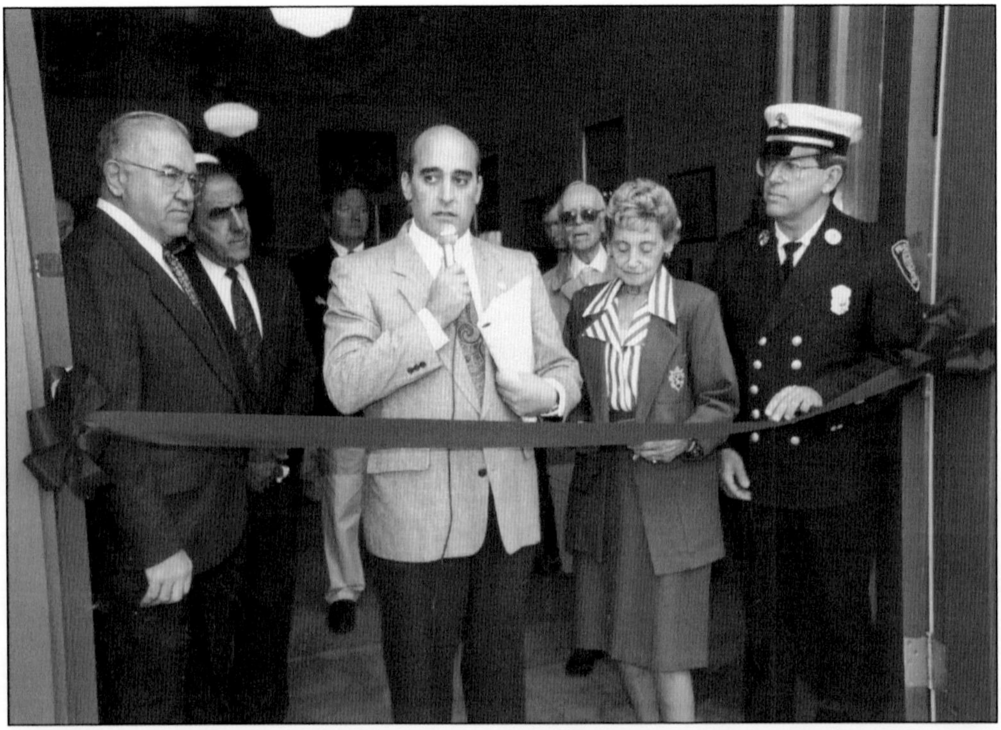

On October 12, 1996, there was much fanfare at the ribbon-cutting ceremony at the new Peabody Fire Museum. From left to right are police chief Joseph Donlon (retired), Mayor Peter Torigian, Joseph Vinchens DiFranco Jr., Rosa Drysdale, and Paul D'Angelo (retired).

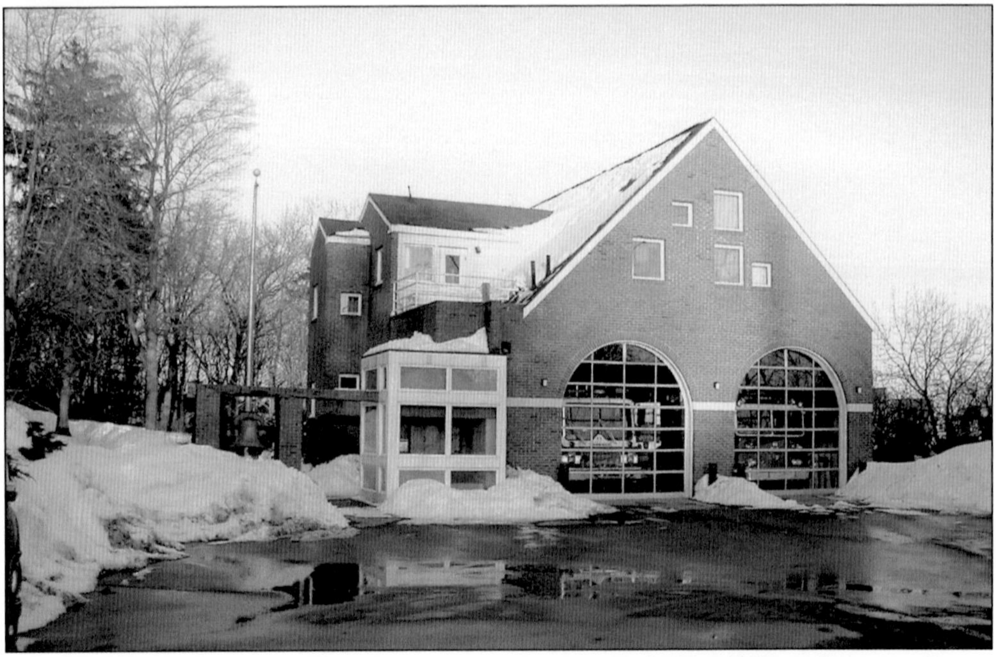

The new engine No. 3's engine house on Prospect Street near the North Shore Shopping Center is a large drive-through, two-bay station. It houses engine No. 3 and ladder No. 2 (reserve).

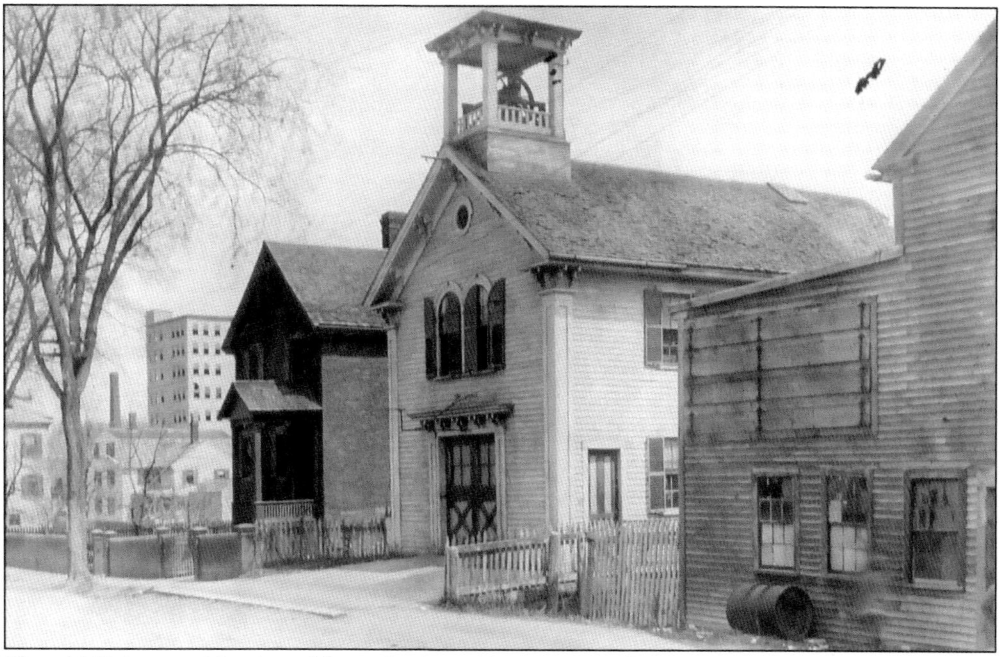

The old Volunteer No. 4, and later the hose No. 4, engine house was on Pierpont Street. This was the second engine house built on this land owned by Gen. William Sutton and was sold by the town around 1900 and moved to its current location on 76 Aborn Street. Hose No. 4 moved to a new engine house on Tremont Street in the east end of town in 1896.

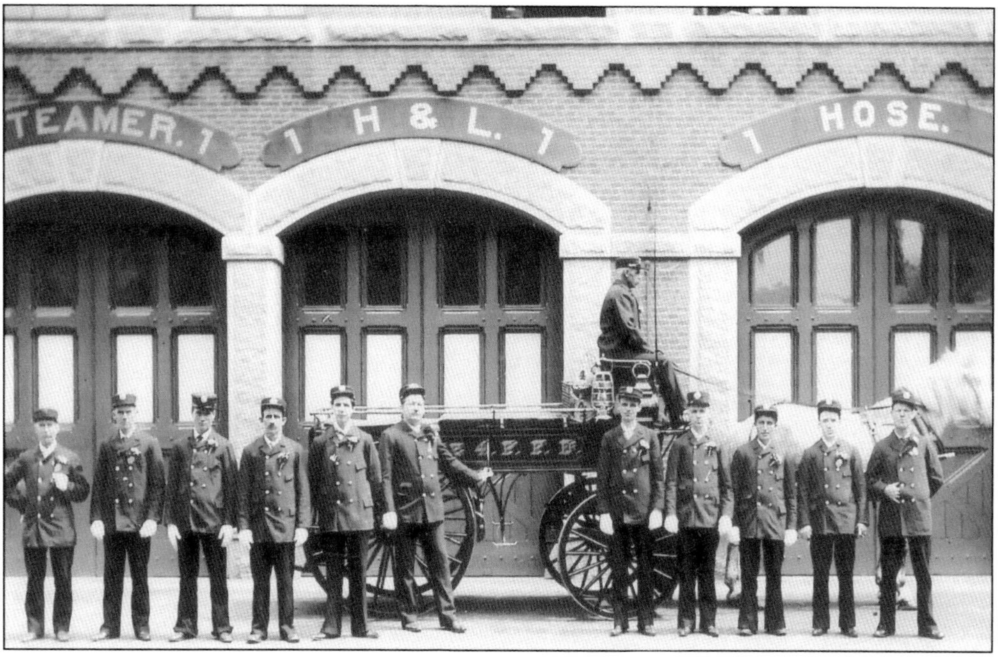

Pictured here is hose No. 4 in front of headquarters. Among those pictured are John Moreland, foreman; Thomas F. Carbrey; William Connolly; Charles Lynch; Thomas P. Masterson; Ulick Brophy, engineer; John Callahan; William Neenan; Bernard Brannigan; Dennis Neenan, and driver John Costello at the reins with Big Jim the horse.

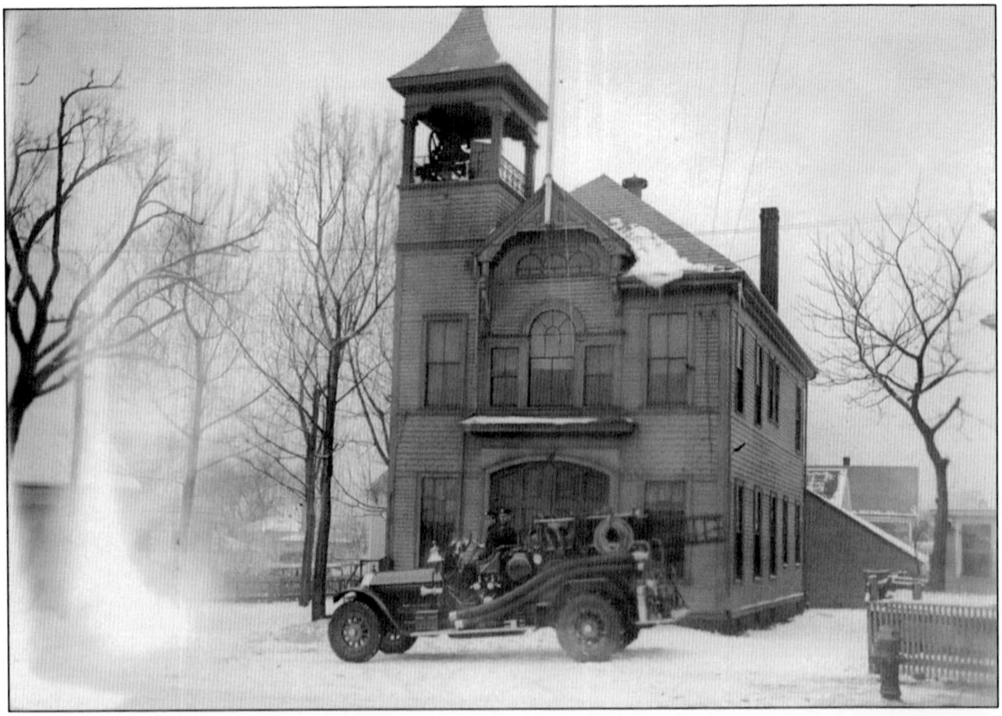

Construction began in 1895 on the new engine house for hose No. 4 at 96 Tremont Street. At a cost of about $7,000, it was completed and an open house was held on March 31 and April 1, 1896. The first horse to have the honor of pulling the hose wagon was named Billy. Big Jim, an iron gray horse, was the second and proved his worth many times over. Engine House No. 4 had a weather vane with a copper flag with the number four cut out of it on its bell tower.

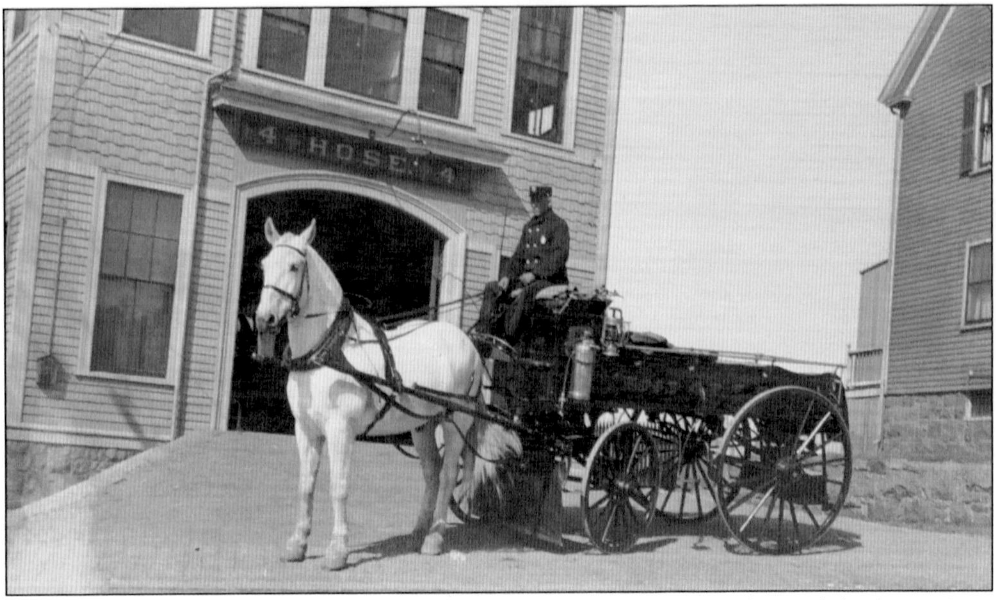

Driver John Costello is seen here with his faithful steed, Big Jim, in front of the hose No. 4 engine house on Tremont Street.

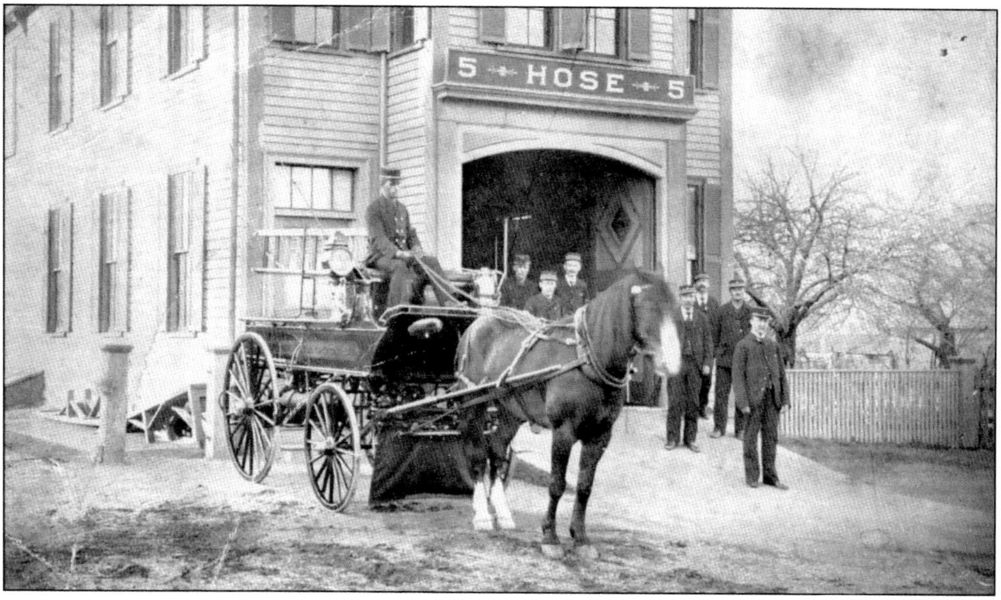

Hose No. 5, built in 1876, is seen here with some of its members in front of their engine house on Lynn Street. The engine house originally housed the Sidney C. Bancroft and later the hose No. 5 wagon. When the station got its first motorized engine in 1916, the horse, whose name was Old Ned, was retired and put out to pasture. The photograph below was taken, in front of Central Station, of hose No. 5 with the company in its dress uniforms and the foreman with its prized trumpet.

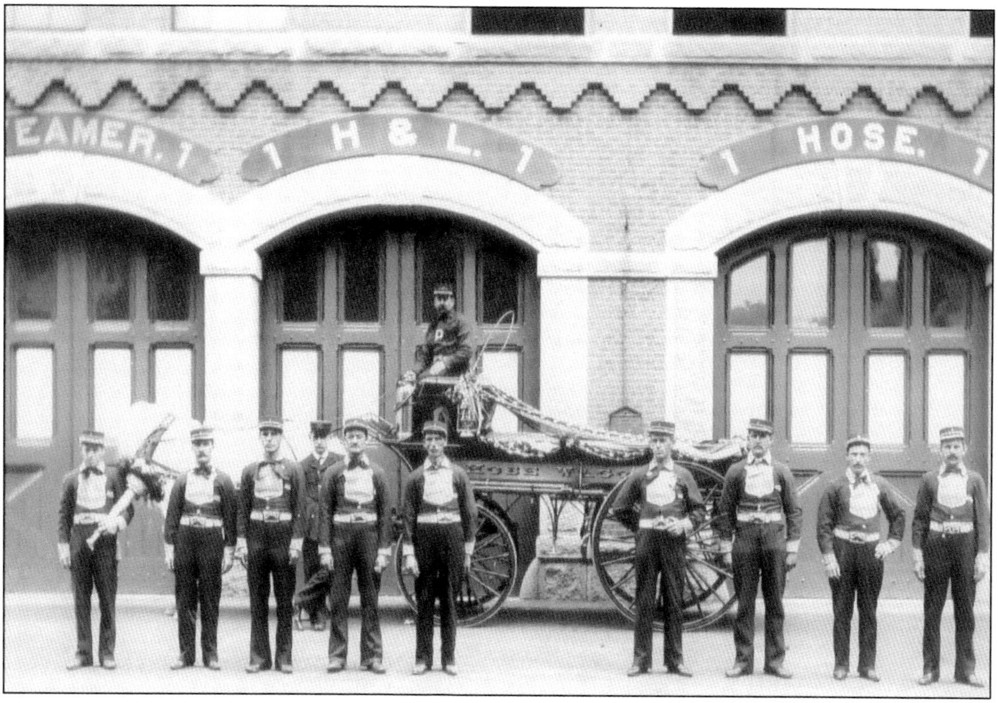

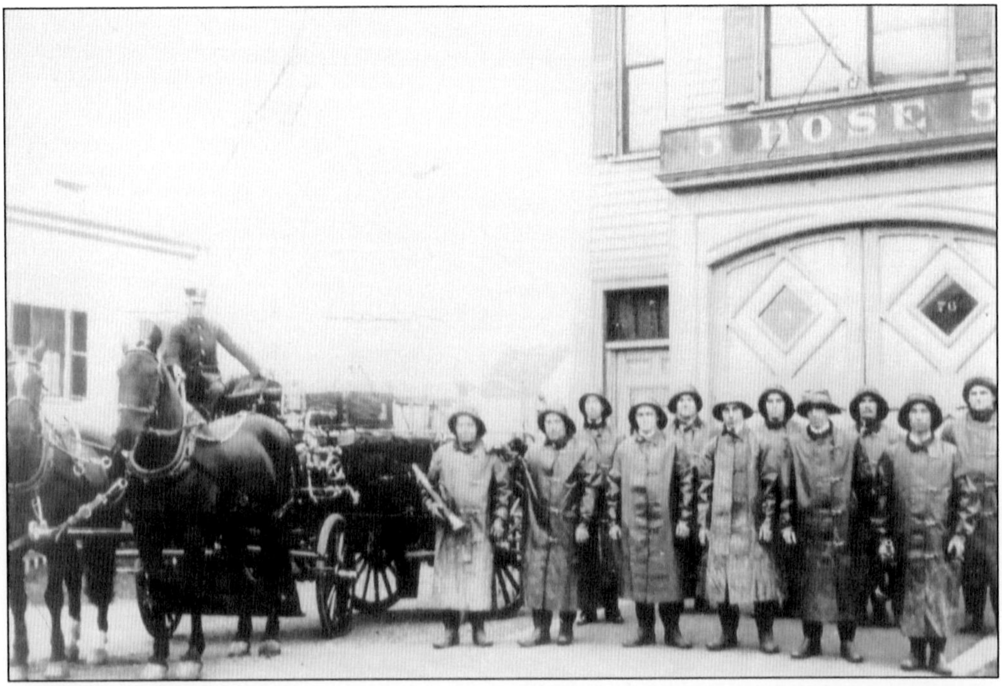

Looking like Gloucester fishermen, the hose No. 5 crew poses out in front of their station with their hose wagon.

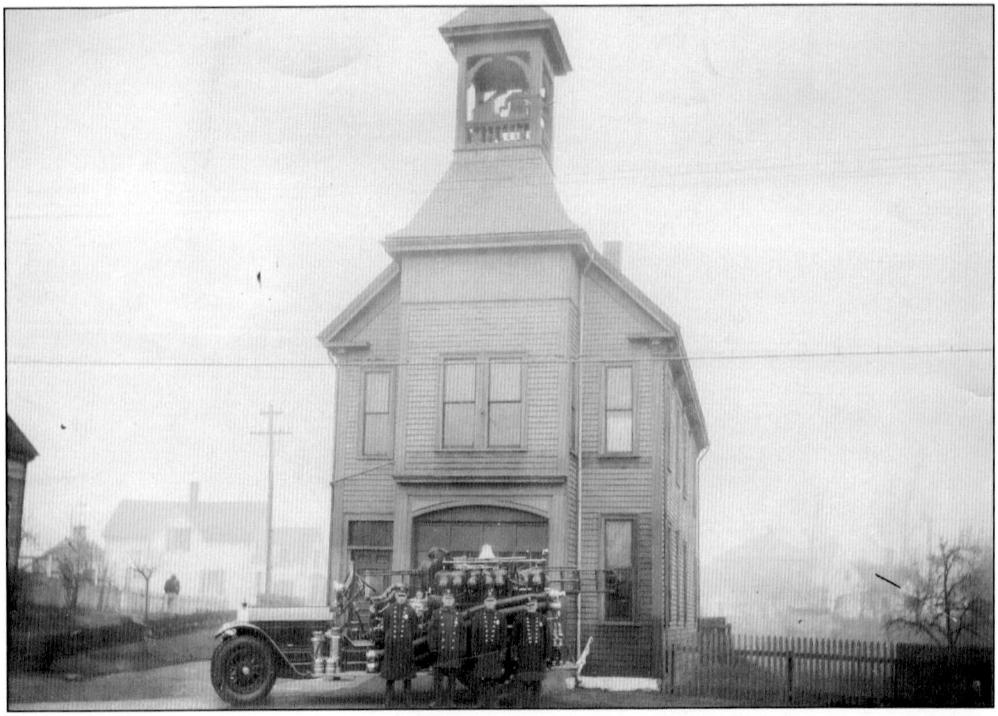

This picture was taken sometime after 1916. It shows the new 1916 LaFrance engine and, from left to right, John McDonald, Francis Kellet, Al Larrabee, and George G. Hodgdon. George G. Hodgdon died on March 1, 1944, while in the performance of his duties.

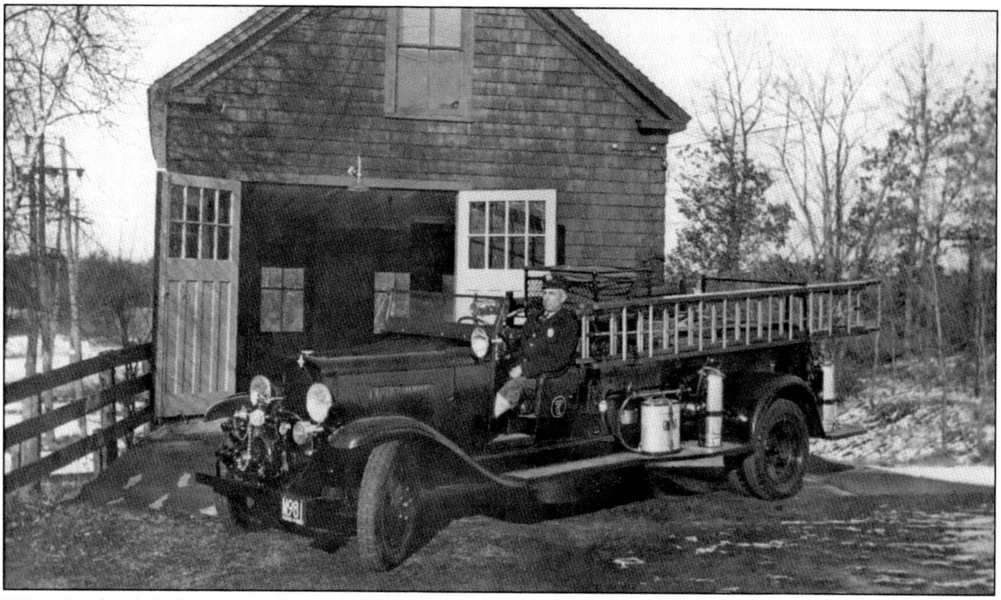

West Peabody's first fire protection came in the form of a used hand engine, the old Niagara. It is not known if a group was formed to man it or not. On November 17, 1899, the Chemical No. 1 Company was formed by George Pauley and Thomas Hutchinson. Its original members consisted of 10 men. In March 1904 at the town meeting, it was voted 82 to 78 to include Chemical No. 1 as members of the Peabody Fire Department. The Chemical No. 1 wagon was housed in different locations, and in 1916, with the taking of land owned by Bryan Healy and the purchase of a motorized engine, a station (pictured above) was built. Also in the picture is engine No. 7, a 1932 Chevrolet chassis engine with a front-mounted, 350-gallon-per-minute Barton pump.

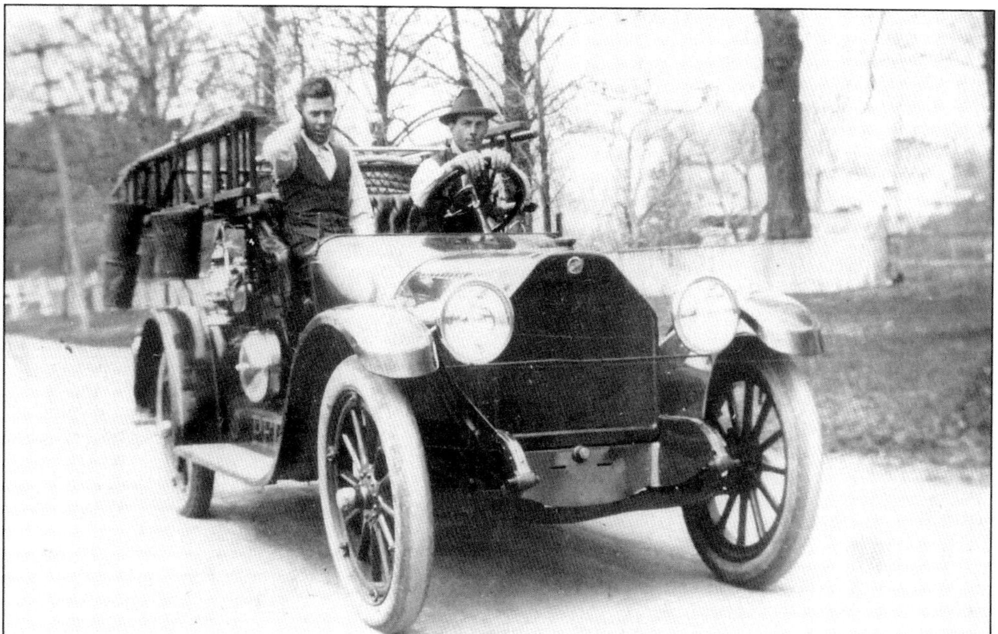

Chemical No. 1, or engine No. 7, from West Peabody is seen here with E. Lawrence Durkee.

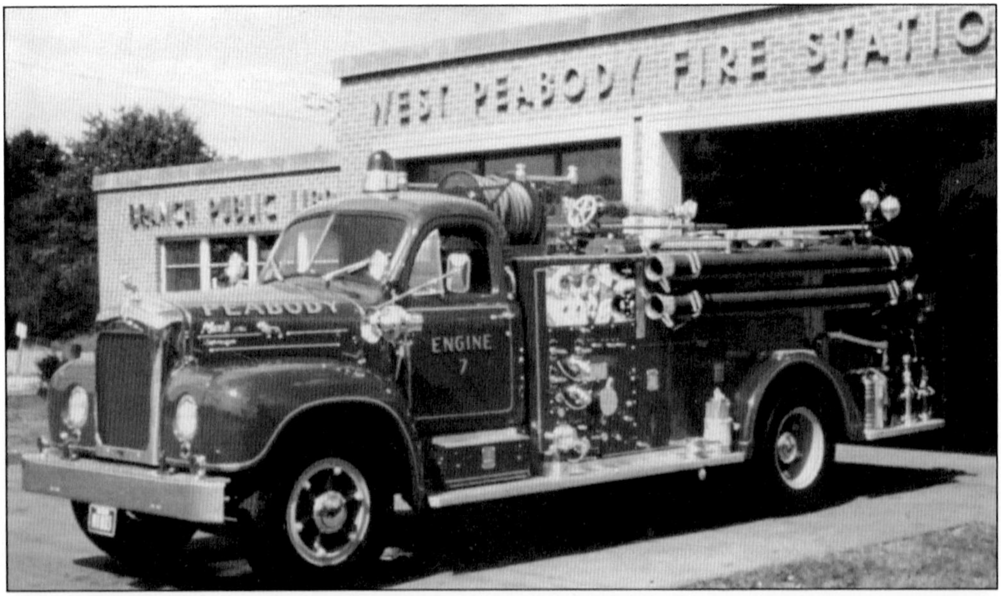

The evolution of West Peabody's fire protection continued with the construction of a new engine house at 597 Lowell Street on land that was purchased by the city from the West Congregational Church. It housed the new engine No. 7, a 1962 Mack B-95 model, and later ladder No. 2, nicknamed "Agnes." The West Peabody branch of the Peabody Institute Library was originally housed on the left side of the building and is now used as quarters for the private ambulance company, North Shore Ambulance, that currently serves Peabody.

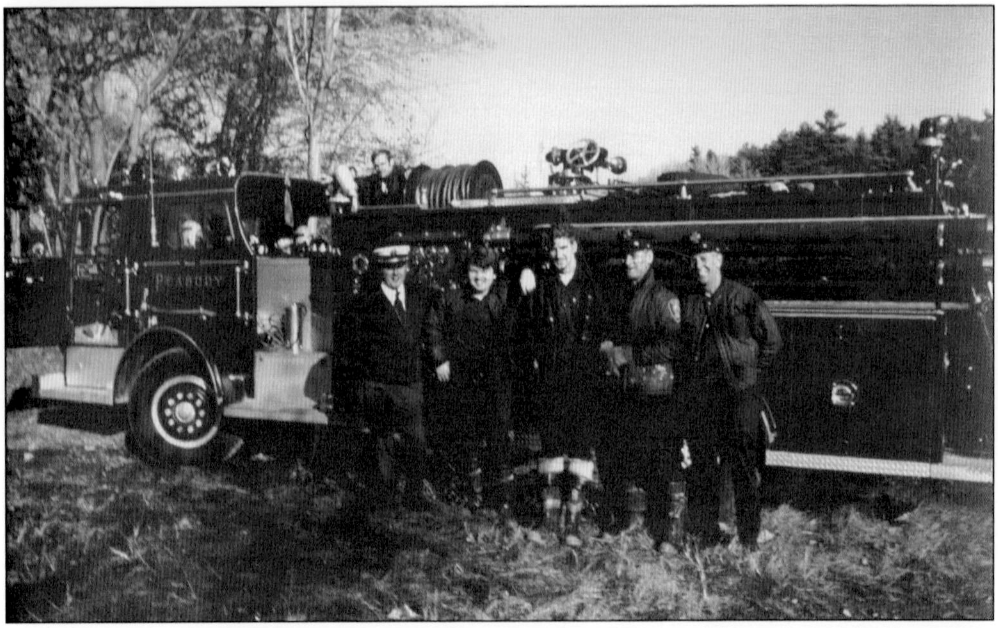

In 1965, the city purchased a new engine for engine No. 1. It was delivered on July 27, 1965. It was a Seagrave Model 800-KB-1000, enclosed cab forwarded, 1,000-gallon-per-minute pump, with a 300-gallon tank. It became engine No. 7 in 1971 and engine No. 4 in 1975. This truck is also pictured in the 1970 North Shore Shopping Center picture (see page 70). It is the one with the side view. It became a reserve piece in 1986 and was junked in 1989.

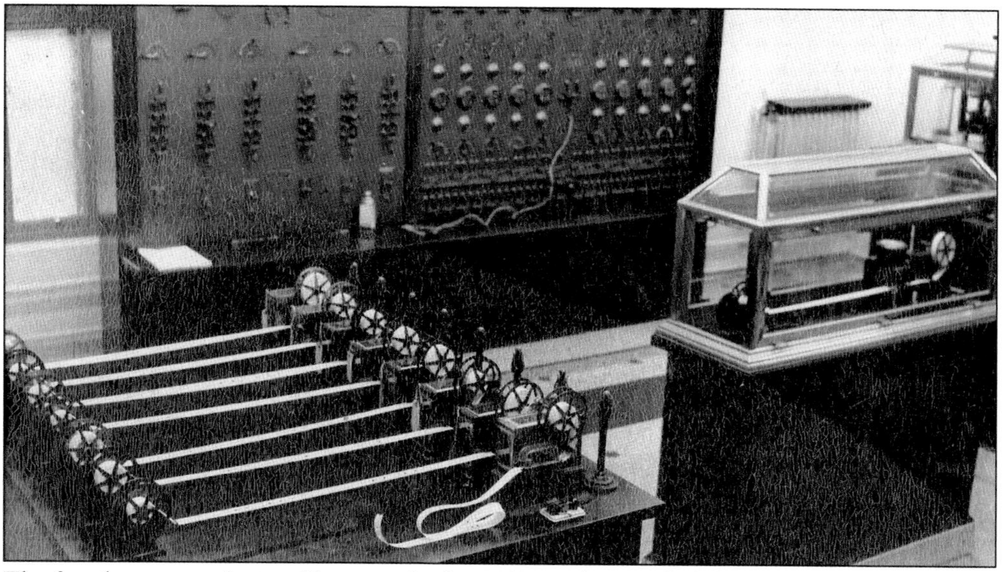

The fire alarm room depicted here was built in 1921 and was called the inner sanctum by many, and only the fire alarm superintendent dared to go in. The first superintendent of the fire alarm was Charles H. Smart. A call man since 1871, he became a permanent man at Central Station on January 1, 1884, and two years later, he was named to fire alarm superintendent with the construction of the new fire alarm system. He retired after the renovation of the fire alarm system. His date of retirement was December 31, 1914, having worked exactly 30 years to the day. With his call man time included, his years of service totaled 43 years.

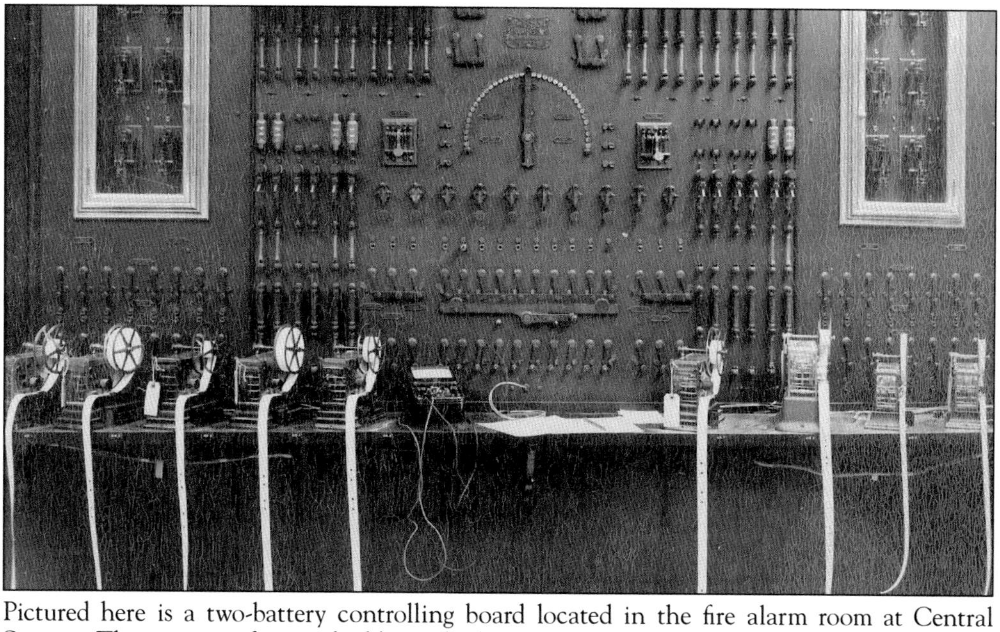

Pictured here is a two-battery controlling board located in the fire alarm room at Central Station. This room, a fireproof addition built in 1921, was located next to and attached to Central Station. It was a Gamewell Fire Alarm, Telephone and Clock system and was considered one of the best around. Its total cost was $20,000. When the room was completed, a series of tests were conducted in July 1921 by the fire underwriter. The underwriters soon after endorsed the system.

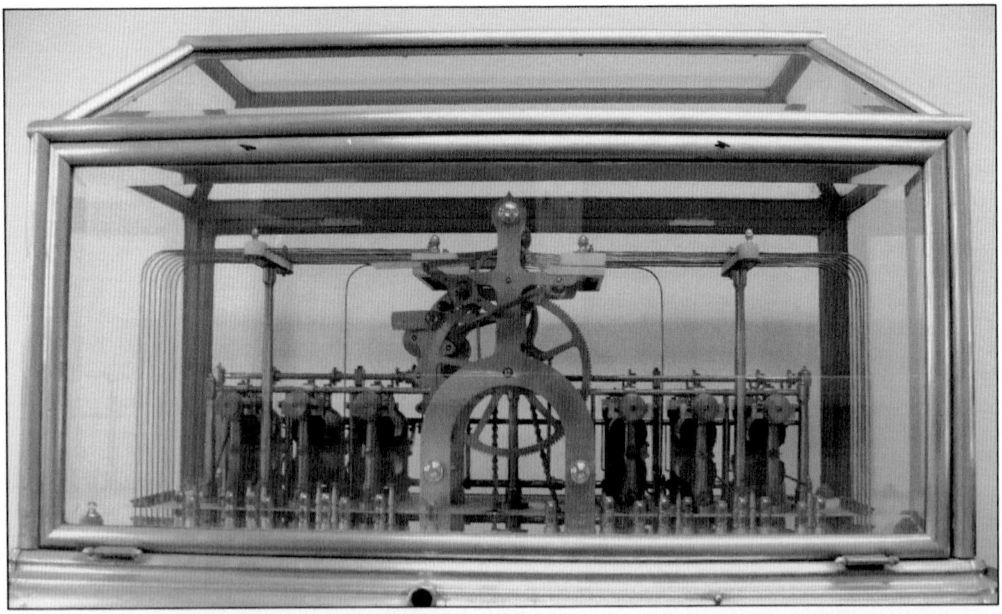

This elaborate glass case of switching relays and circuitry, property of the Peabody Historical Society, is called a repeater. Once a fire alarm pull box is activated, the telegraph circuit opens and closes, ringing the bell or horns, and records on spools of paper tape a number that coincides with the box, giving the location of the activated fire alarm box.

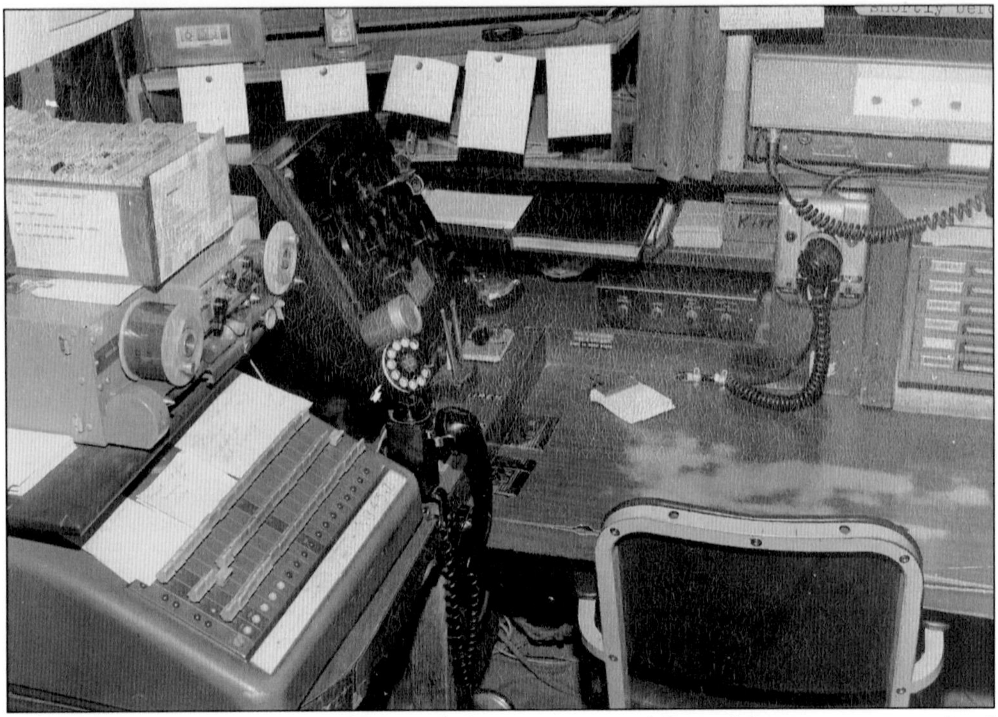

This desk was on the apparatus floor until the renovations of Central Station were completed in 1989.

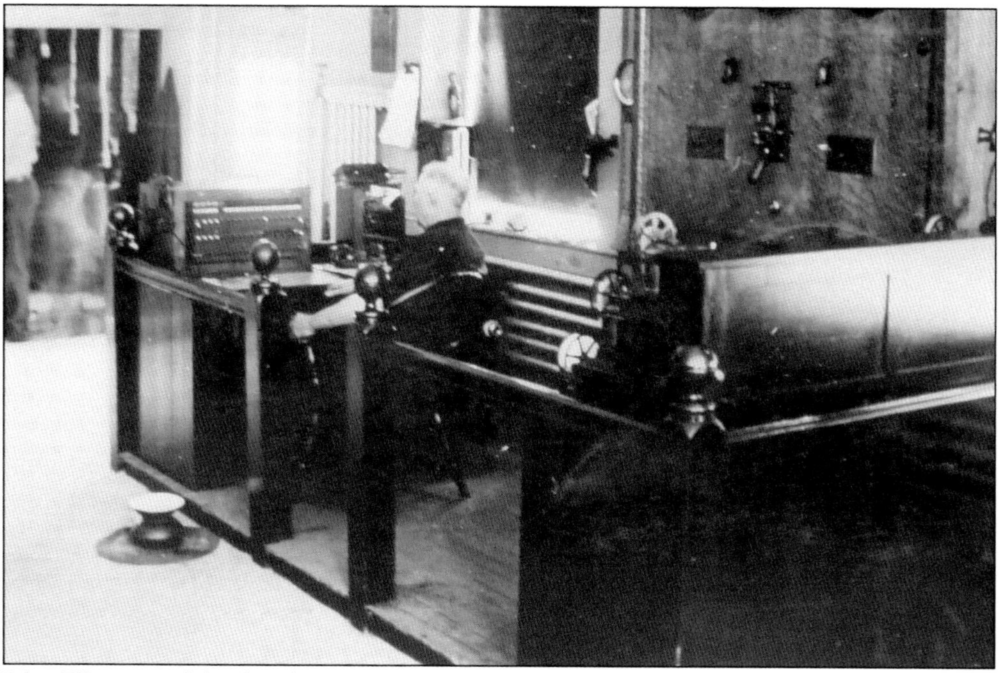

John Ellis, one of the fire alarm operators, is shown here at his desk that he watched over for many years. This watch desk was located out on the apparatus floor and was connected to the fire alarm room.

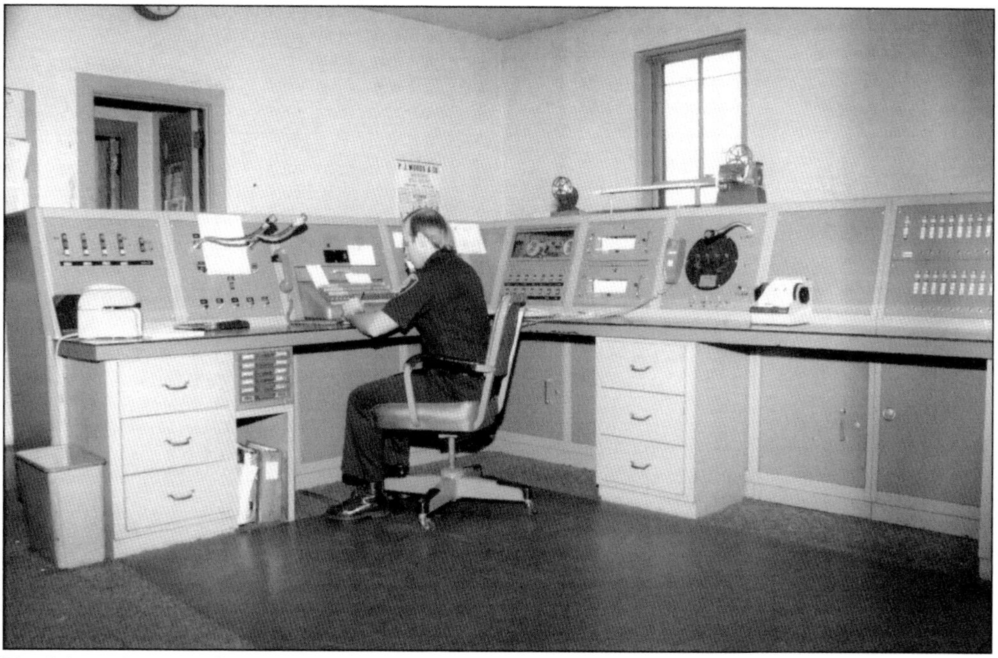

Firefighter Greg Koulas, now a captain at Engine No. 7 in West Peabody, is shown here working the new fire alarm. The room was built in 1921 and renovated. It is pictured here in 1988, and not many changes occurred during those years. The old slate circuit boards had to be smashed in order to get them out.

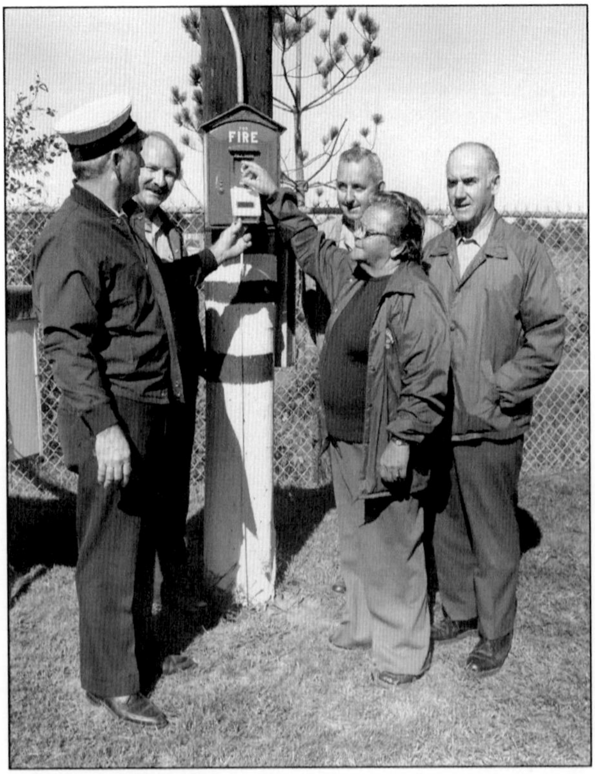

This fire alarm pull box was installed after one of the trailer parks on Route 1 was split in two due to the construction of Interstate 95. Seen here, from left to right, are Deputy Chief Frank Connors, Mike Eliuk, Roger Fecteau, trailer park activist Liz Warden, and Buddy Leonard.

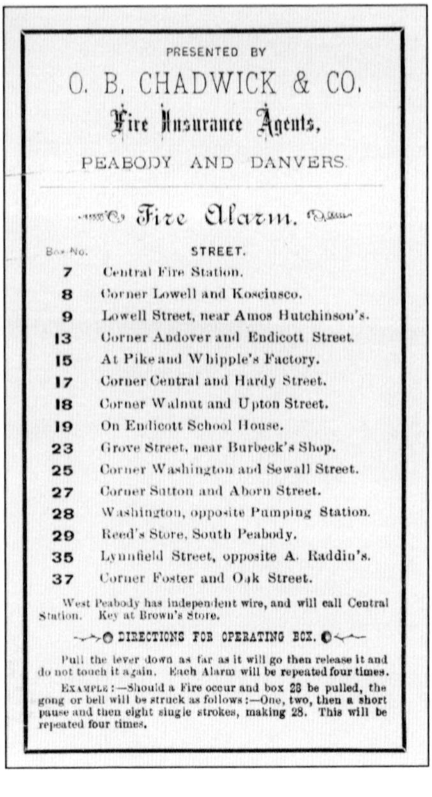

PRESENTED BY

O. B. CHADWICK & CO.

Fire Insurance Agents,

PEABODY AND DANVERS.

Fire Alarm.

Box No.	STREET.
7	Central Fire Station.
8	Corner Lowell and Kosciusco.
9	Lowell Street, near Amos Hutchinson's.
13	Corner Andover and Endicott Street.
15	At Pike and Whipple's Factory.
17	Corner Central and Hardy Street.
18	Corner Walnut and Upton Street.
19	On Endicott School House.
23	Grove Street, near Burbeck's Shop.
25	Corner Washington and Sewall Street.
27	Corner Sutton and Aborn Street.
28	Washington, opposite Pumping Station.
29	Reed's Store, South Peabody.
35	Lynnfield Street, opposite A. Raddin's.
37	Corner Foster and Oak Street.

West Peabody has independent wire, and will call Central Station. Key at Brown's Store.

DIRECTIONS FOR OPERATING BOX.

Pull the lever down as far as it will go then release it and do not touch it again. Each Alarm will be repeated four times.

EXAMPLE:—Should a Fire occur and box 28 be pulled, the gong or bell will be struck as follows:—One, two, then a short pause and then eight single strokes, making 28. This will be repeated four times.

Fire alarm cards like this one were passed out by insurance companies giving the list of the fire alarm boxes in the city. This card is from O. B. Chadwick and Company. The initial installation of the fire alarm system was in 1886, and Charles H. Smart was named the first superintendent. It was a Gamewell system with 16 street boxes, a whistle, and a tower bell on Central Station, all installed at a cost of $3,500. The first alarm by this system was box 17, on August 1, 1886, at 2:10 a.m. for a barn fire.

Three

HORSES AND THEIR APPARATUS

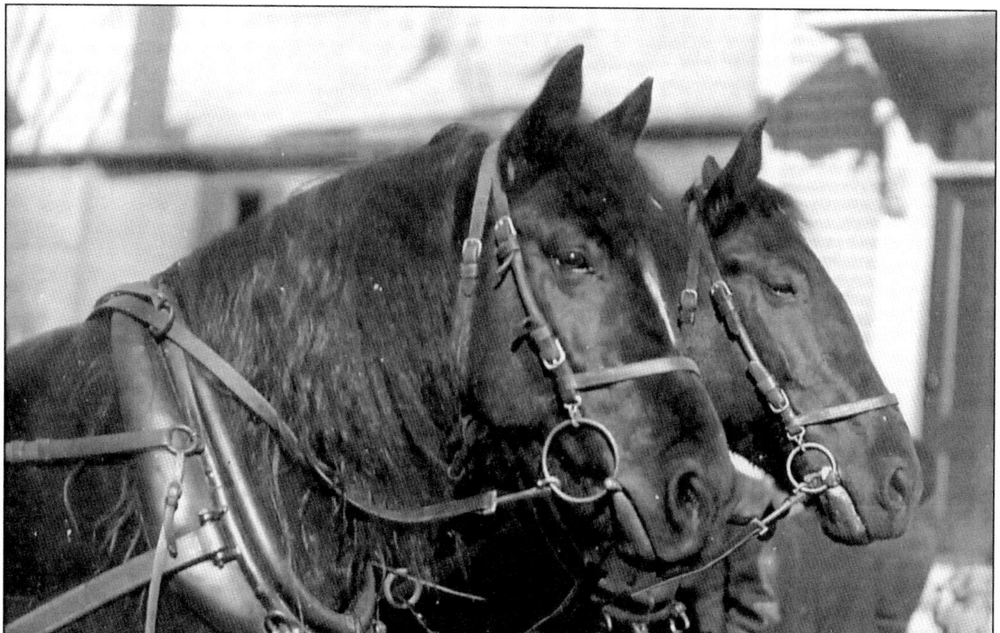

With names like Tom and Jerry, Jumbo, Old Ned, Big Jim, and Billy, horses in the fire service were considered to be members of the department. If they did not show up or were sick, the firemen would have to pull the equipment to the scene, which was nearly impossible. So they were treated very well. Jumbo was one of the horses located at Central Station. He was put down because of sickness, and the newspapers wrote an article about the sadness the firefighters endured from their loss. Tom and Jerry were the last two horses to be used in the Peabody Fire Department.

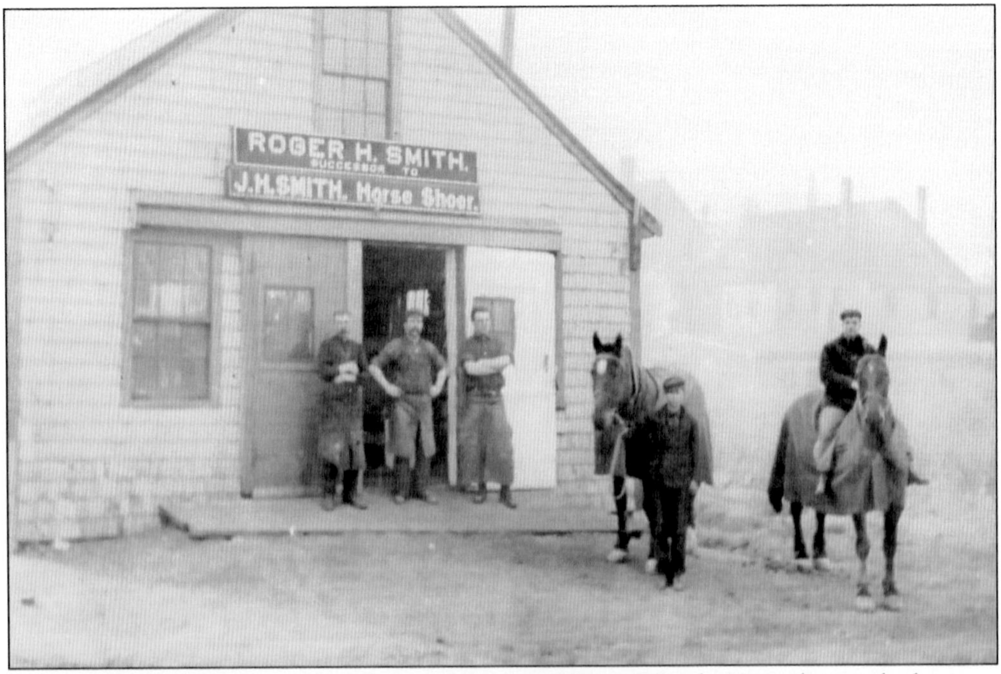

Pictured here is the blacksmith and horseshoe shop of Roger Smith. He took over the business from Jesse H. Smith, who was a blacksmith and horseshoer. Jesse was also on the fire department in Peabody.

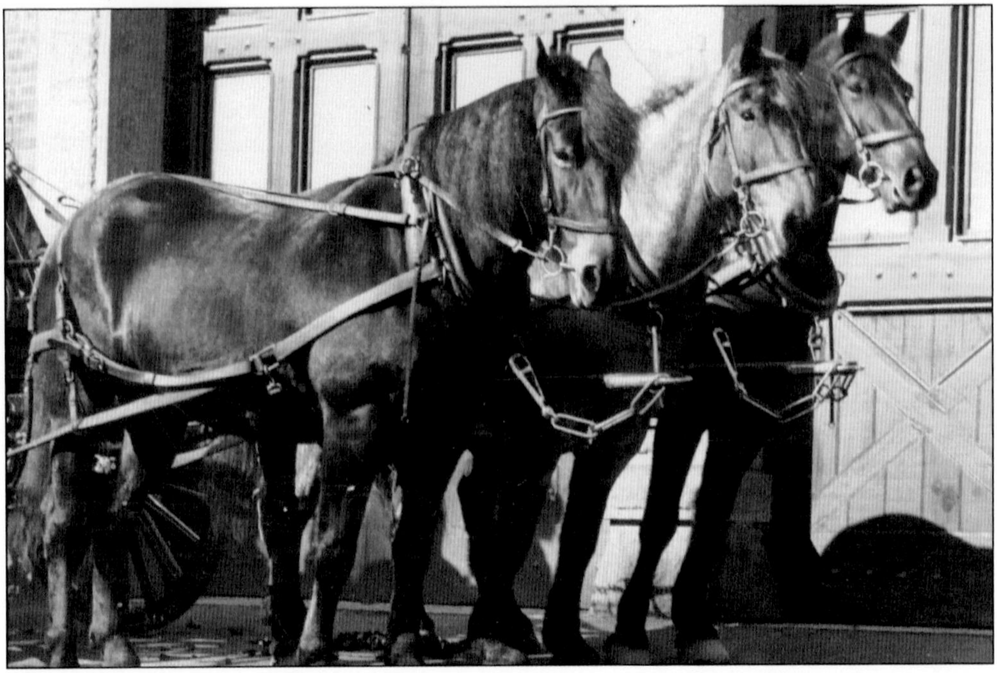

When horses were first used, they were brought to the engine houses from private citizens and other city departments, such as the street department, and hitched up. As time went on and there was a greater need for permanent horses, the city started to purchase horses for fire department purposes only.

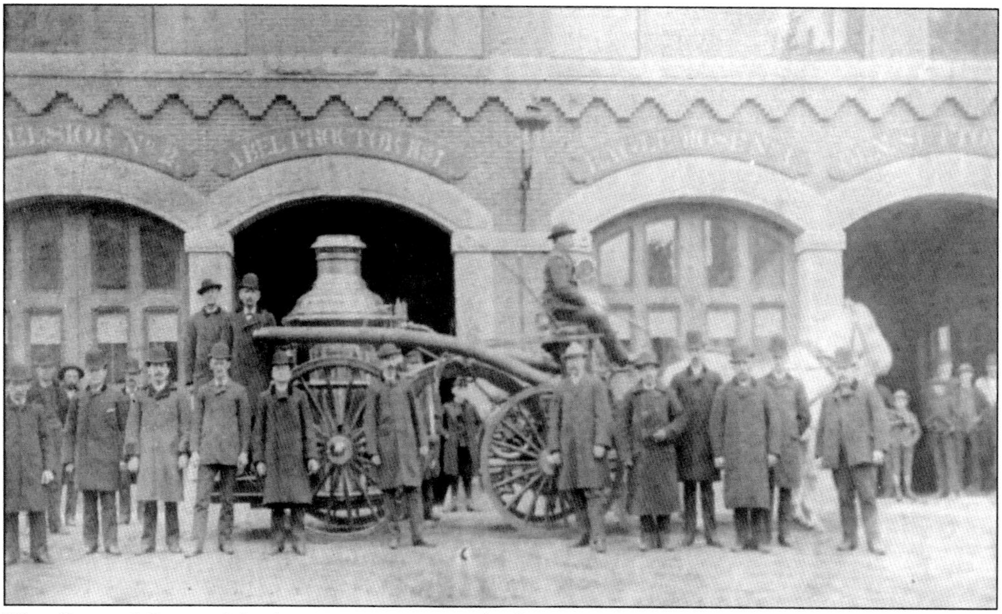

The General Sutton No. 1 steam engine was built in 1865 and was originally housed on Foster Street until 1873 when Central Station was built. It is shown here with some of its crew in front of Central Station. It was the town's first steamer engine, and it was built by the L. Button Company of Waterford, New York. It arrived on April 22, 1865. In 1877, the town purchased a new General Sutton steamer from the LaFrance Company.

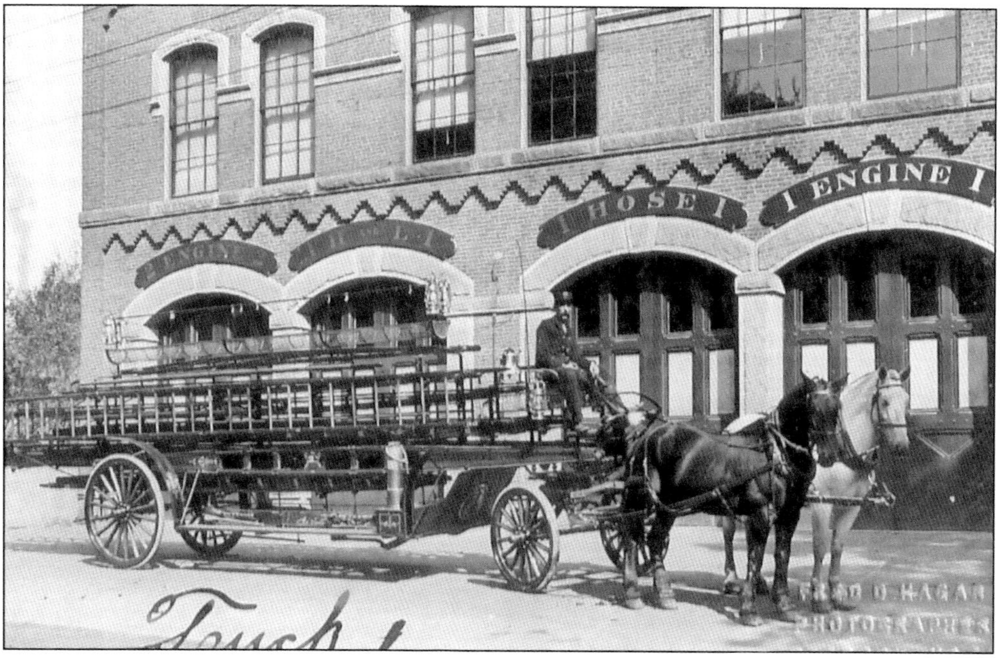

The personal names attached to the apparatus dropped off at some point, and they were just named according to the type of vehicle that they were. Seen here, truck No. 1 with a two-horse hitch indicates it was an earlier horse-drawn ladder truck. The new, heavier ones needed three-horse hitches.

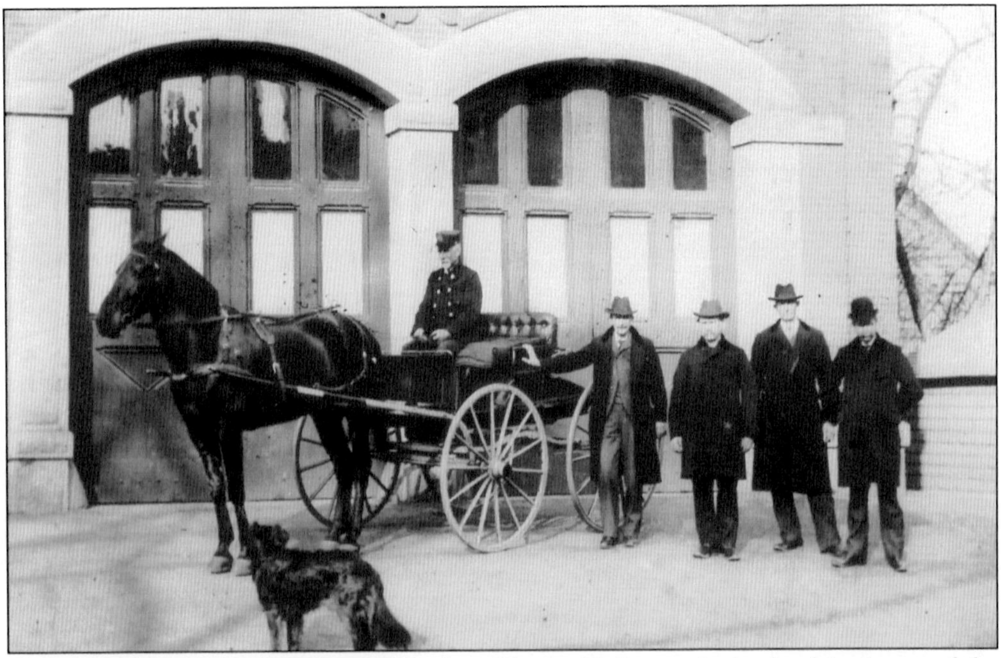

As the chief's aide, fire alarm superintendent Charles H. Smart sits in the driver's seat of the chief's buggy. This wagon was used for the chief engineer of the board of engineers and the fire alarm superintendent. The other men pictured here are probably the chief engineer and his board.

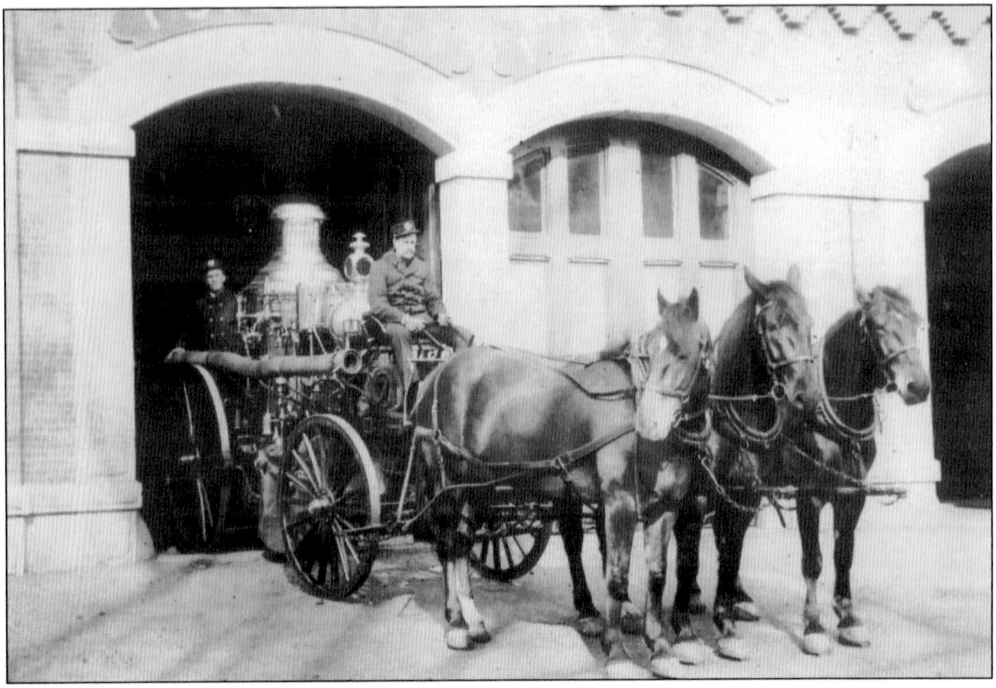

Steamer No. 1 is pictured here with its three-horse hitch and driver.

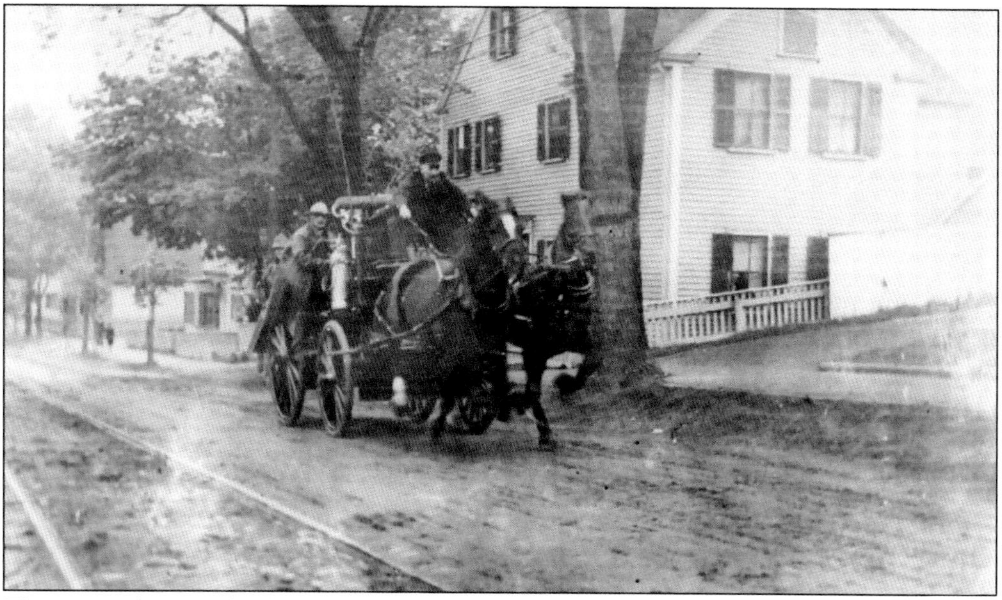

These two photographs show the combination hose wagons in action. The photograph above shows the wagon heading west on Lowell Street, and the photograph below shows the wagon heading south on Foster Street. Notice the feet of the horses; it appears all four hooves are off the ground.

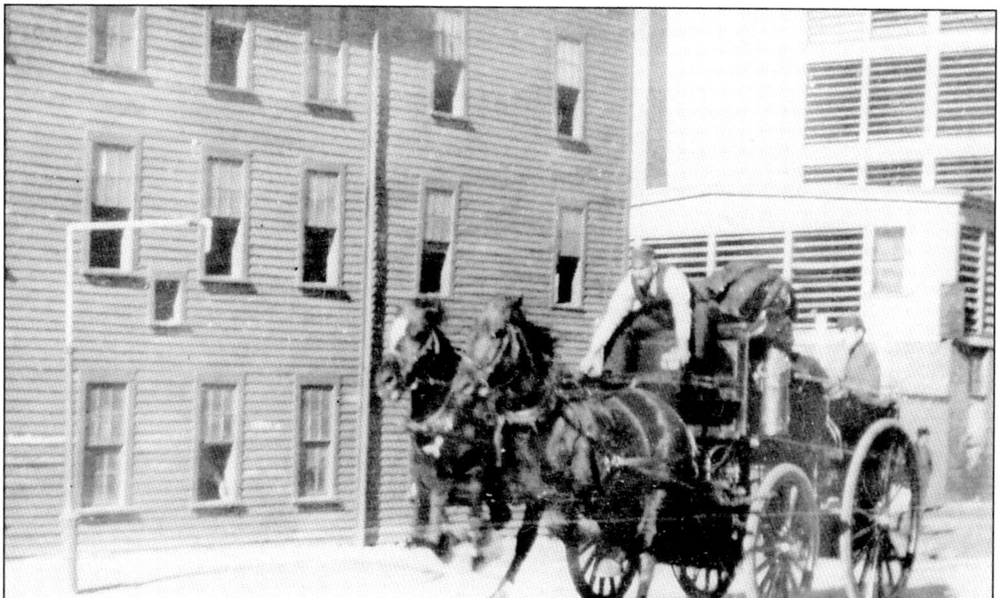

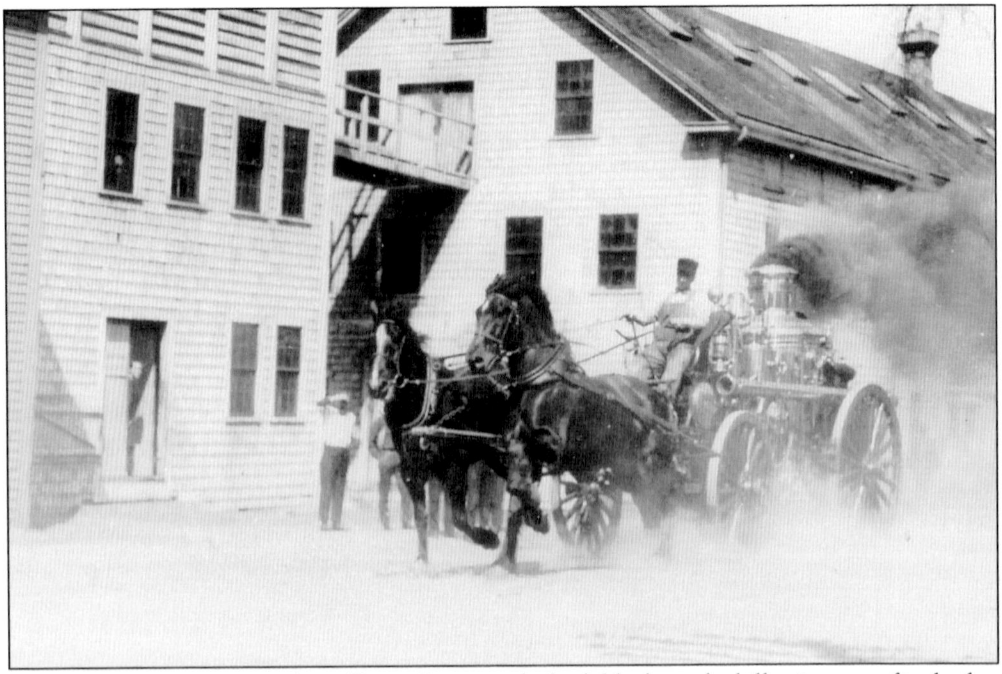

Steamer No. 1, seen racing down Foster Street with thick black smoke billowing out of its boiler, is ready to put water on the fire when it gets to the scene.

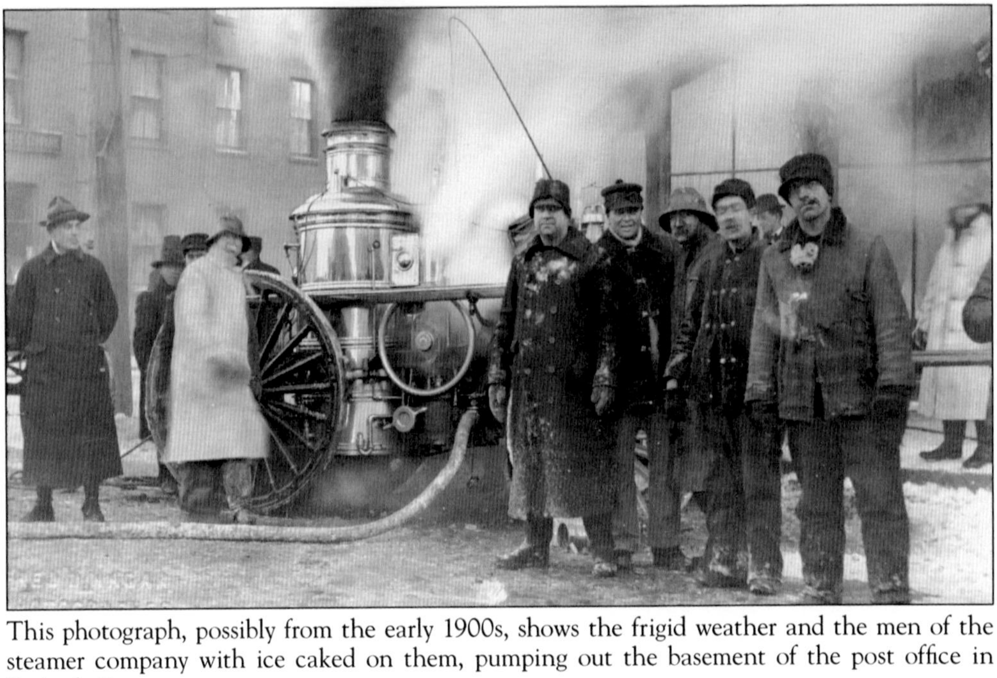

This photograph, possibly from the early 1900s, shows the frigid weather and the men of the steamer company with ice caked on them, pumping out the basement of the post office in Peabody Square.

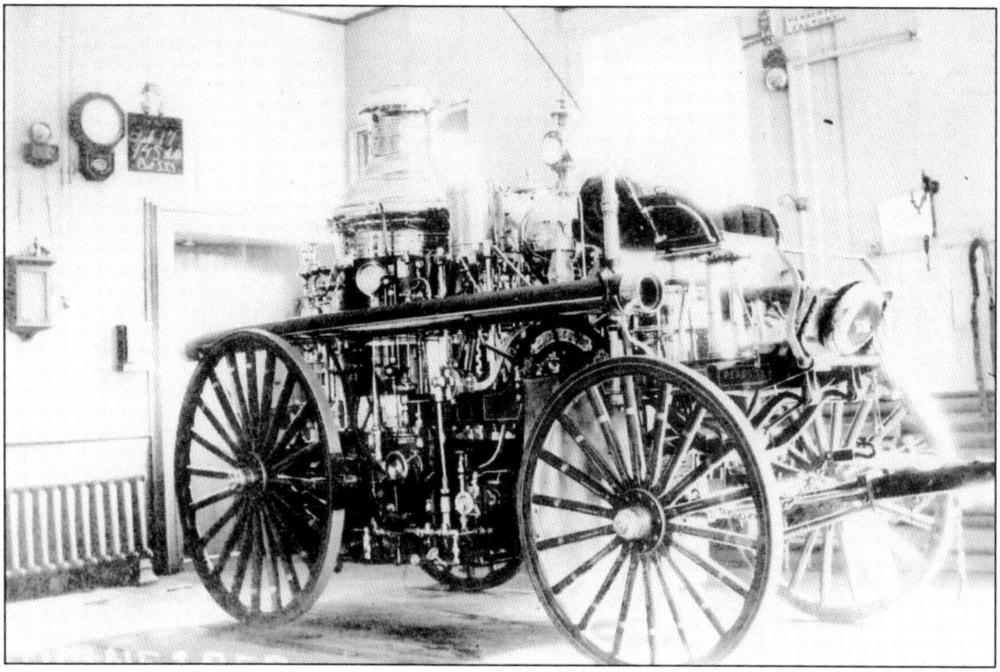

Steamer engine No. 1, ready for action, waits patiently in its bay for the bell to go off. Notice in front of the driver's seat is a large gong or bell that would alert pedestrians of the oncoming steamer, just like the sirens of today.

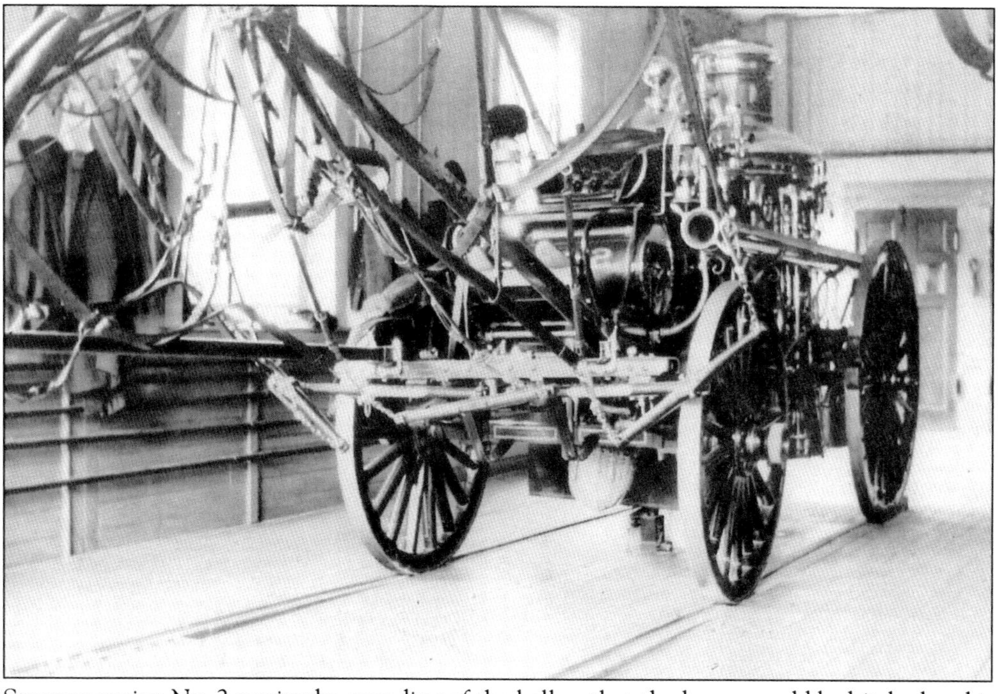

Steamer engine No. 2 awaits the sounding of the bell so that the horses could be hitched and it could be on its way to help put out a fire.

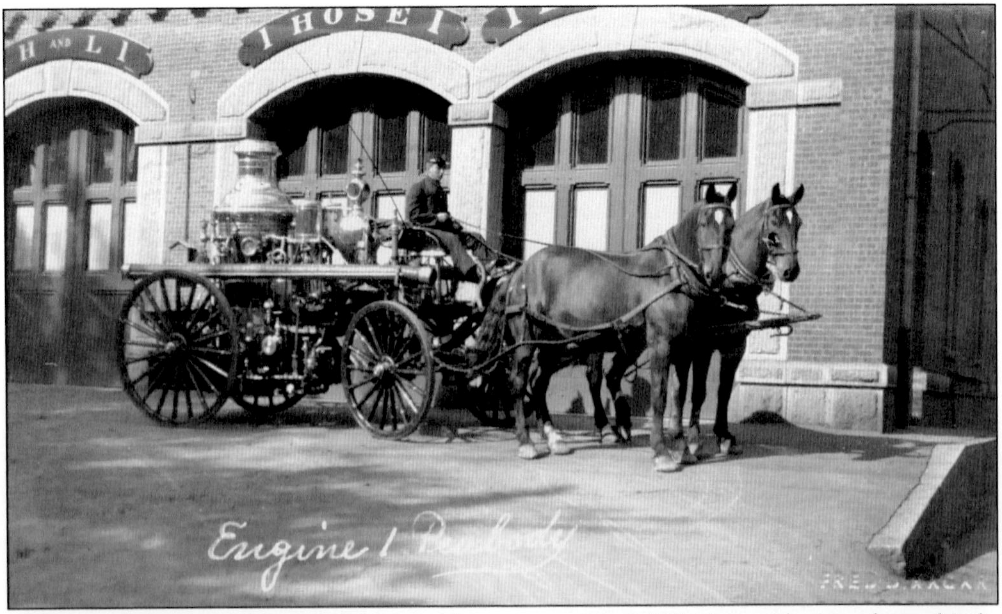

The picture above is of steamer No. 1 on the ramp at Central Station with a two-horse hitch. The picture below shows steamer No. 2, which was bought later and was heavier and used a three-horse hitch. The second steam engine that the town purchased was a L. Button steam fire engine called the Excelsior, and it arrived on July 10, 1874. The word *excelsior* means "curled wood chips," which is what was used with oil as a kindling for the coal to power the steam engines. The Excelsior, along with a hose wagon, cost $4,760, and the hose, bought from Josiah Gates and Sons, cost $612.37. The engine was retired in March 1902 when the new, larger Amoskeag steamer arrived.

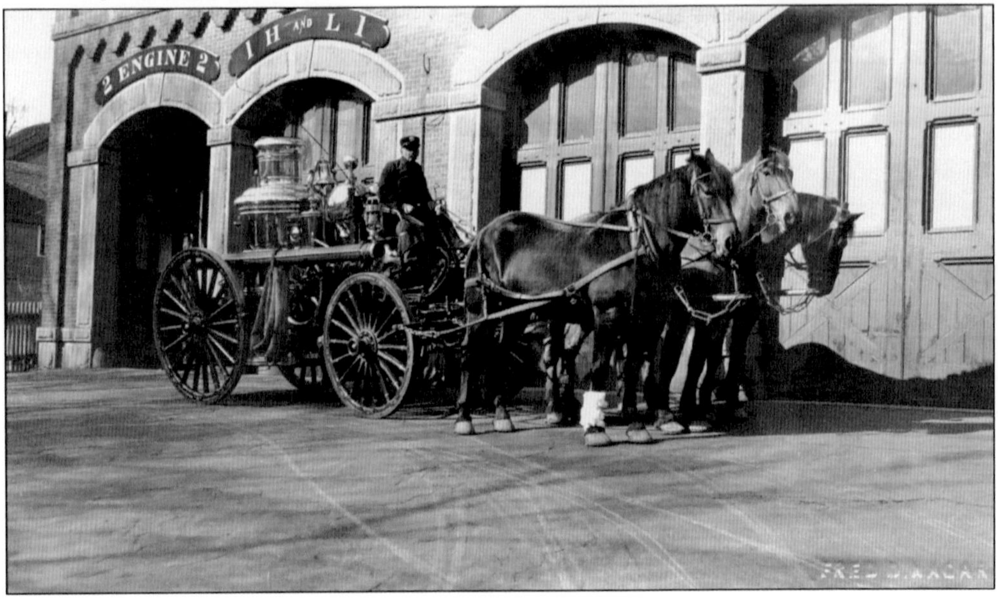

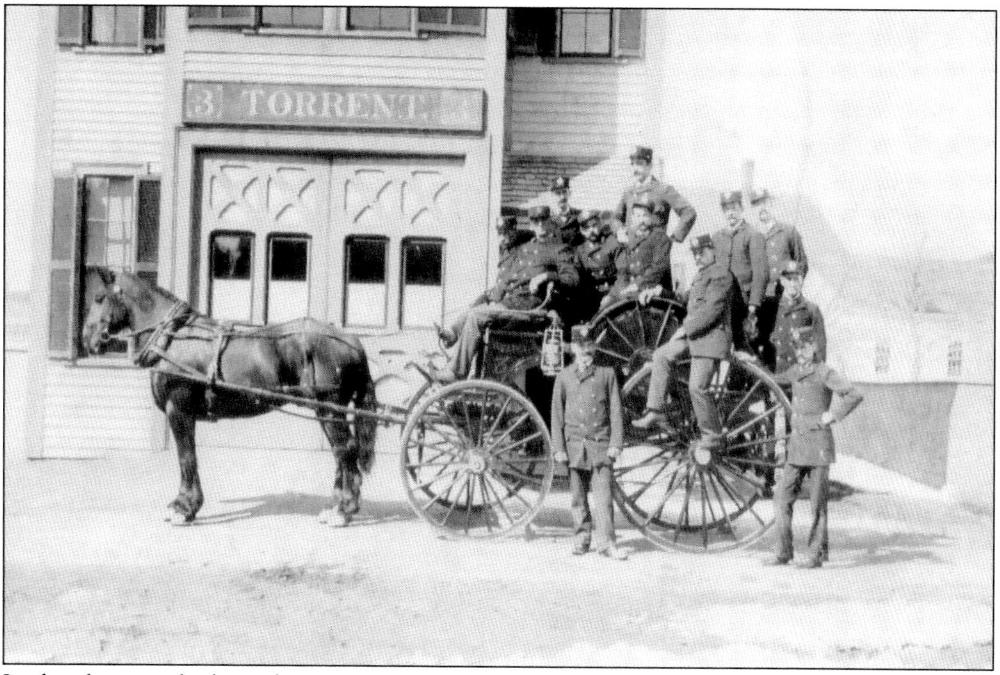

In the photograph above, hose carriage No. 3, with some of its men, is positioned in front of its quarters, near Wilson Square. Wilson Square was once called Pine Tree Corner. Hose wagon No. 3 replaced hose carriage No. 3 and is shown below with some of its men. From left to right are Freeman Bodge (driver), Benjamin Bodge, Herbert Buxton, John Cotton, George Bodge, David Spence, George Morgan, and John Toye.

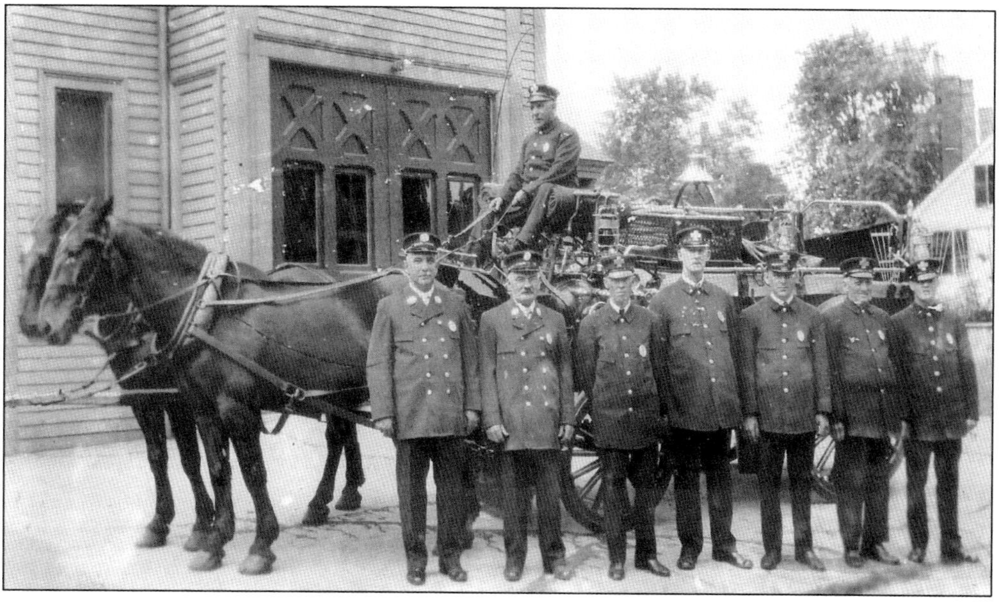

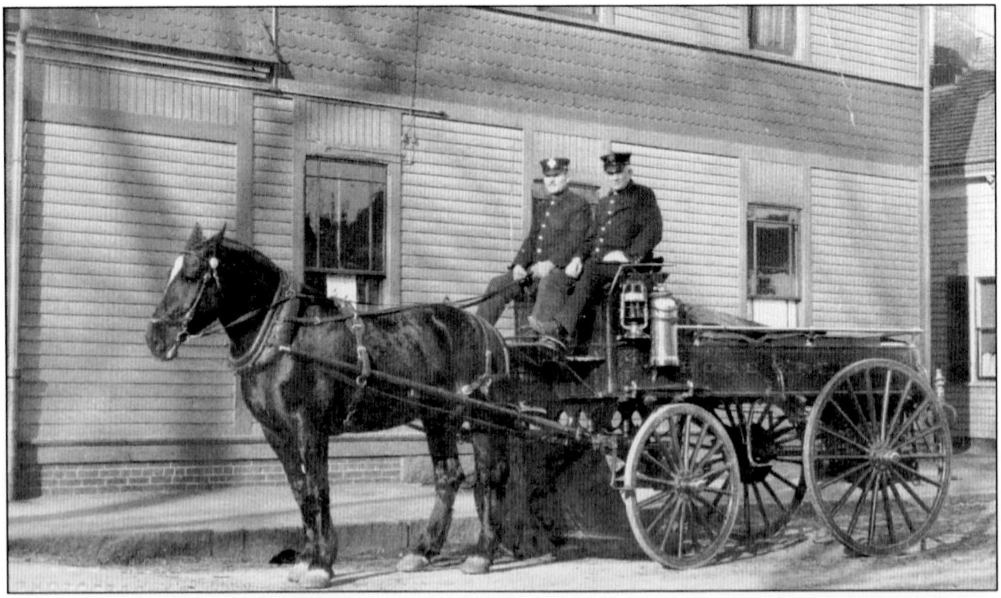

The hose wagons brought the hose to the fire. Pictured above is hose wagon No. 2, which was located at the corner of Washington and Mason Streets. Hose wagon No. 5, pictured below, was located in the southern part of town.

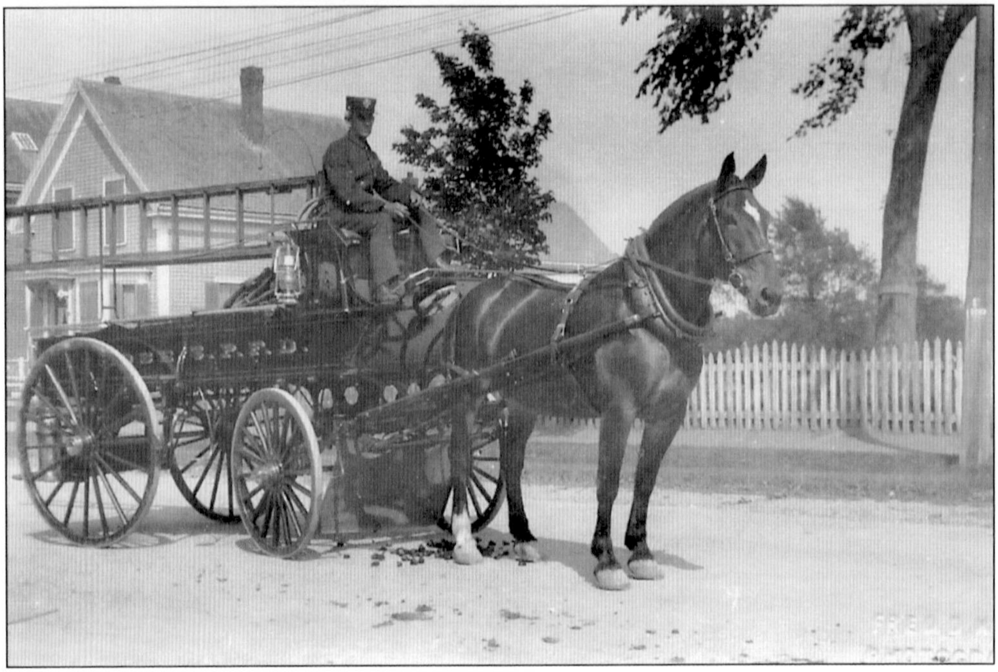

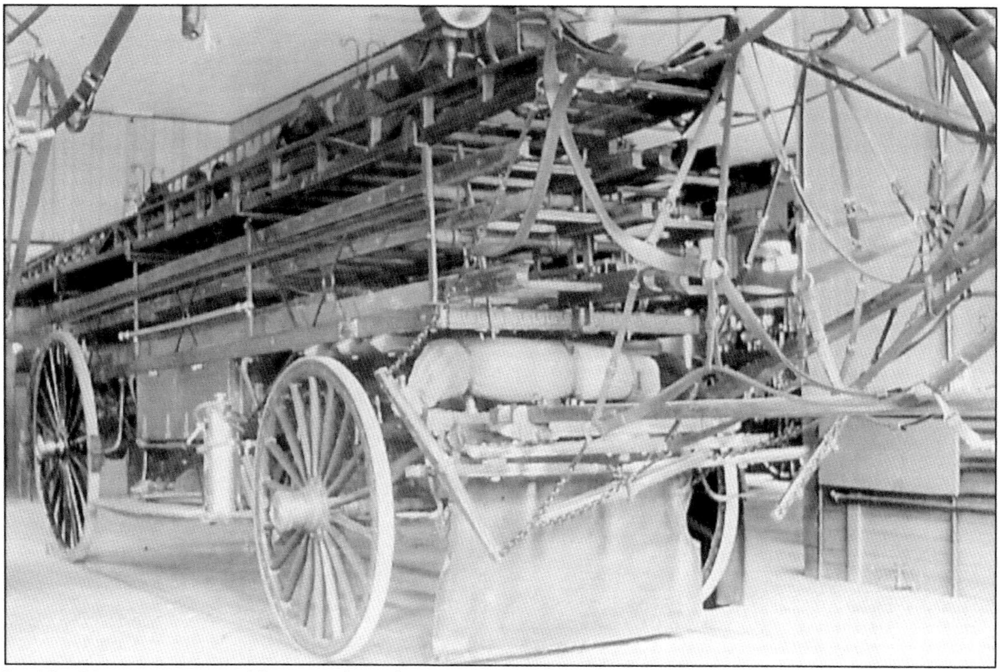

Ladder No. 1 is pictured here with its harness dangling in front of the truck, awaiting the horses to be hitched up and to be pulled to the fire.

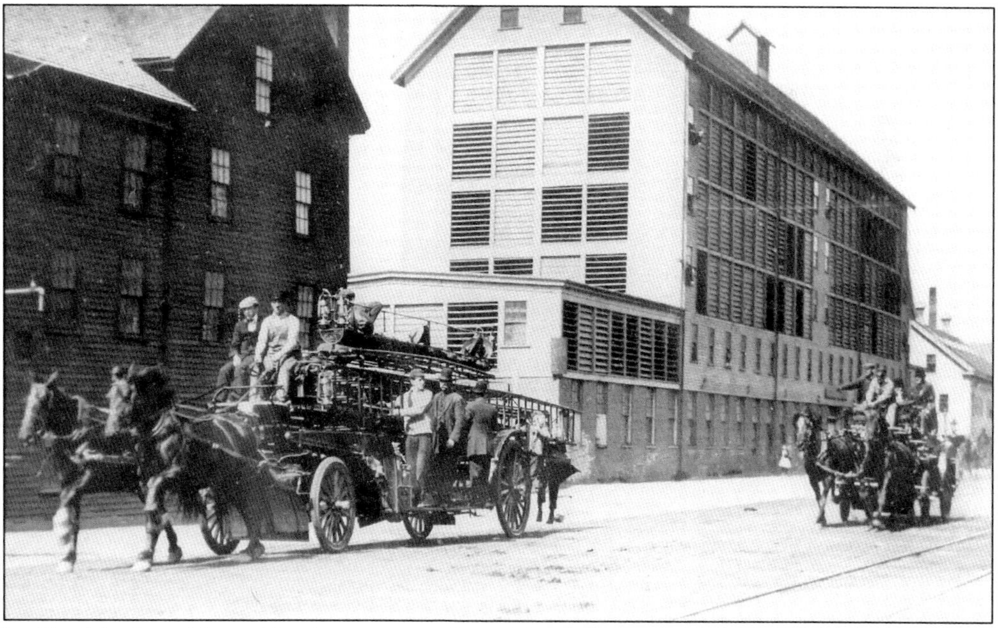

Ladder No. 1 and hose No. 1 are racing down Foster Street, near Mason Street, in front of Southwick Leather Company. Across the street was a photographer's shop, and many pictures were taken there.

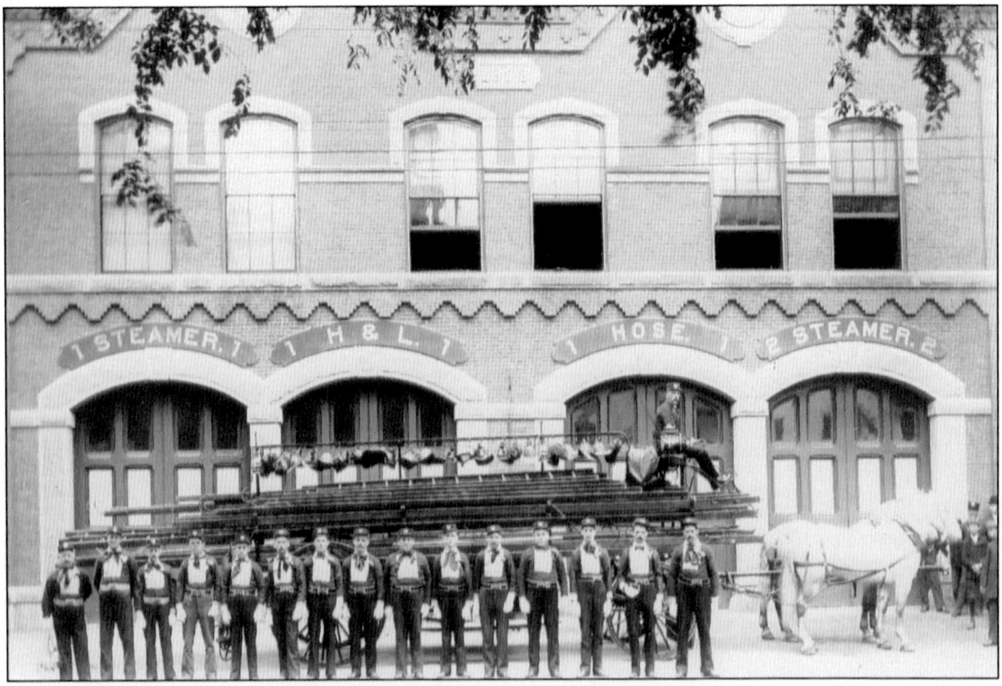

Hook and ladder No. 1 is seen above with its crew displaying its apparatus, horses, and new uniforms with pride. Ladder No. 1, with horses in full gallop, is seen below responding to a fire. The ladder truck would slow down at the square, and the firemen would jump onto the truck. Notice the policeman riding the sideboard.

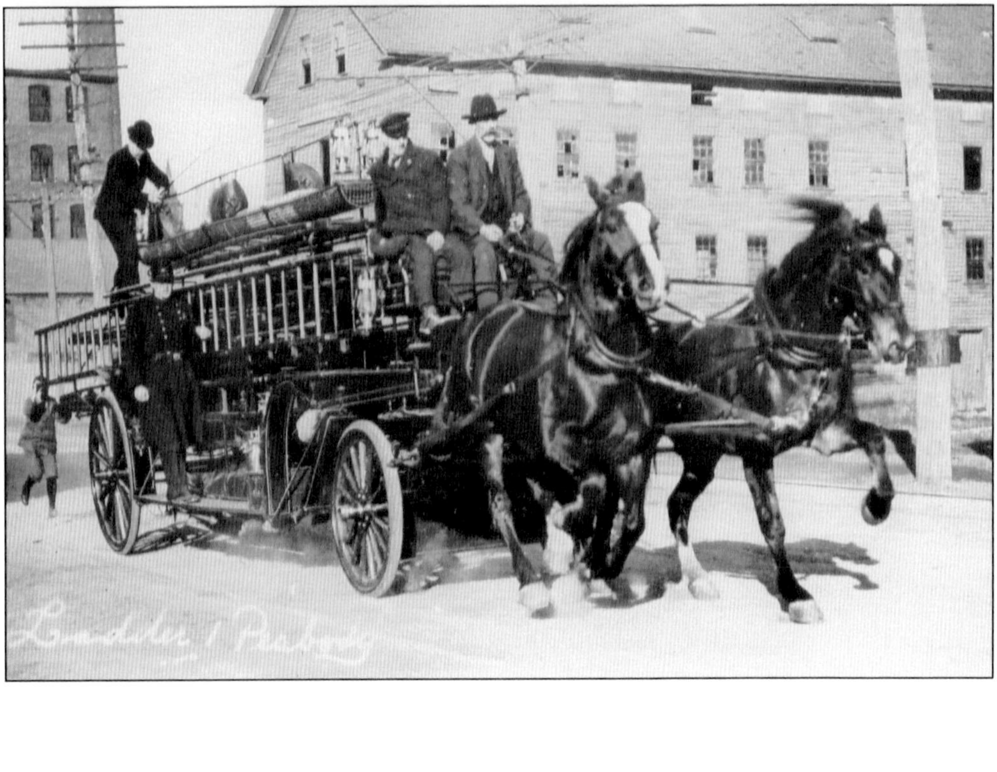

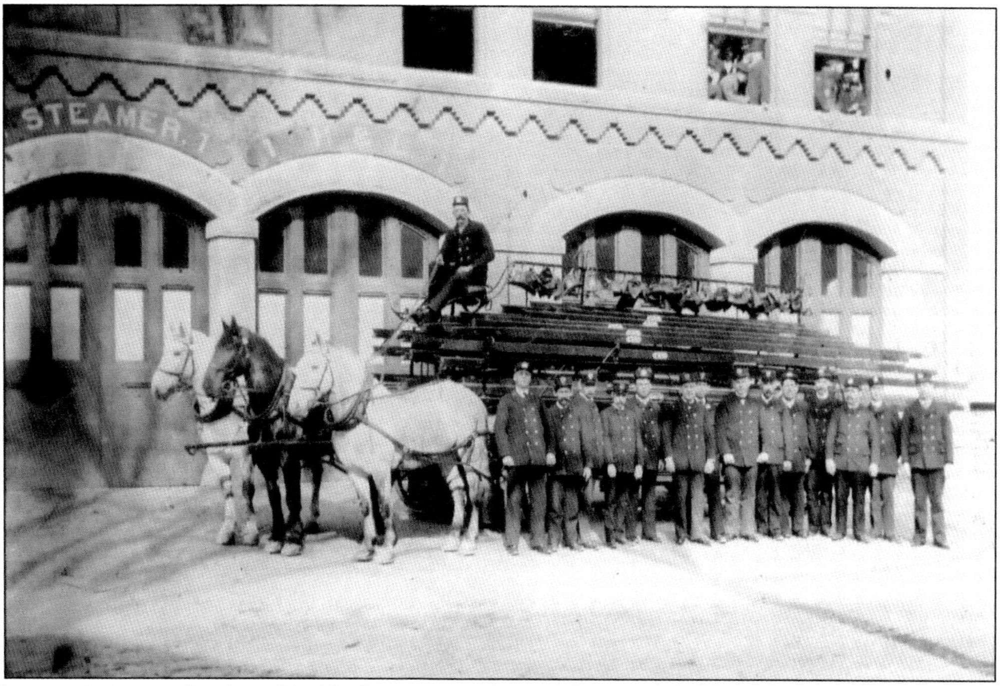

Pictured above is hook and ladder No. 1, seen here with its crew and with people looking down from a second floor, and below is ladder No. 1 stopping for a picture on the side of the road.

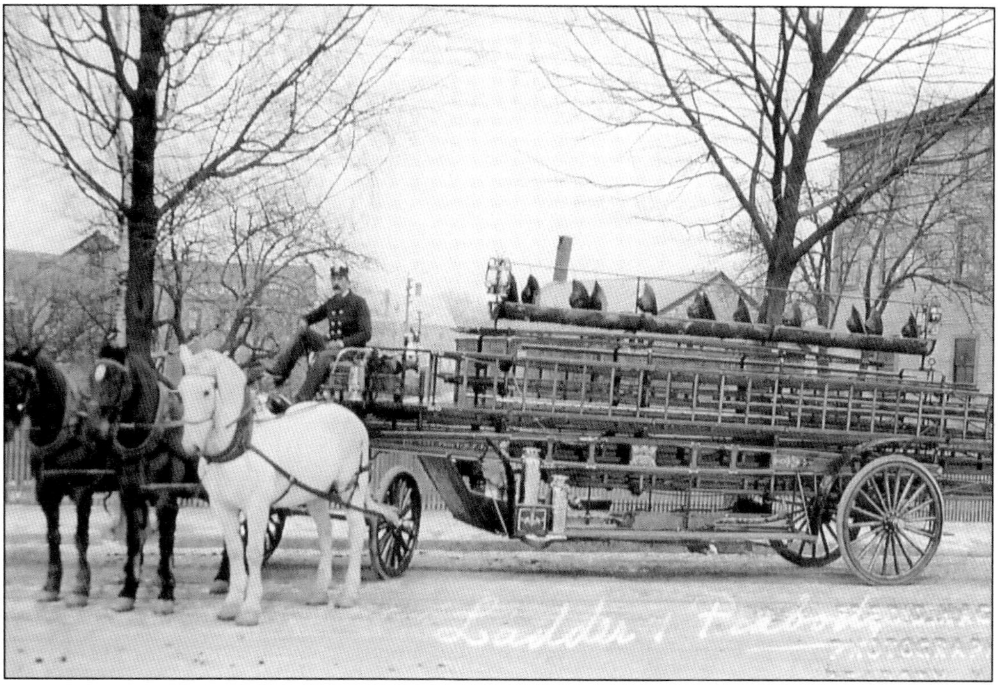

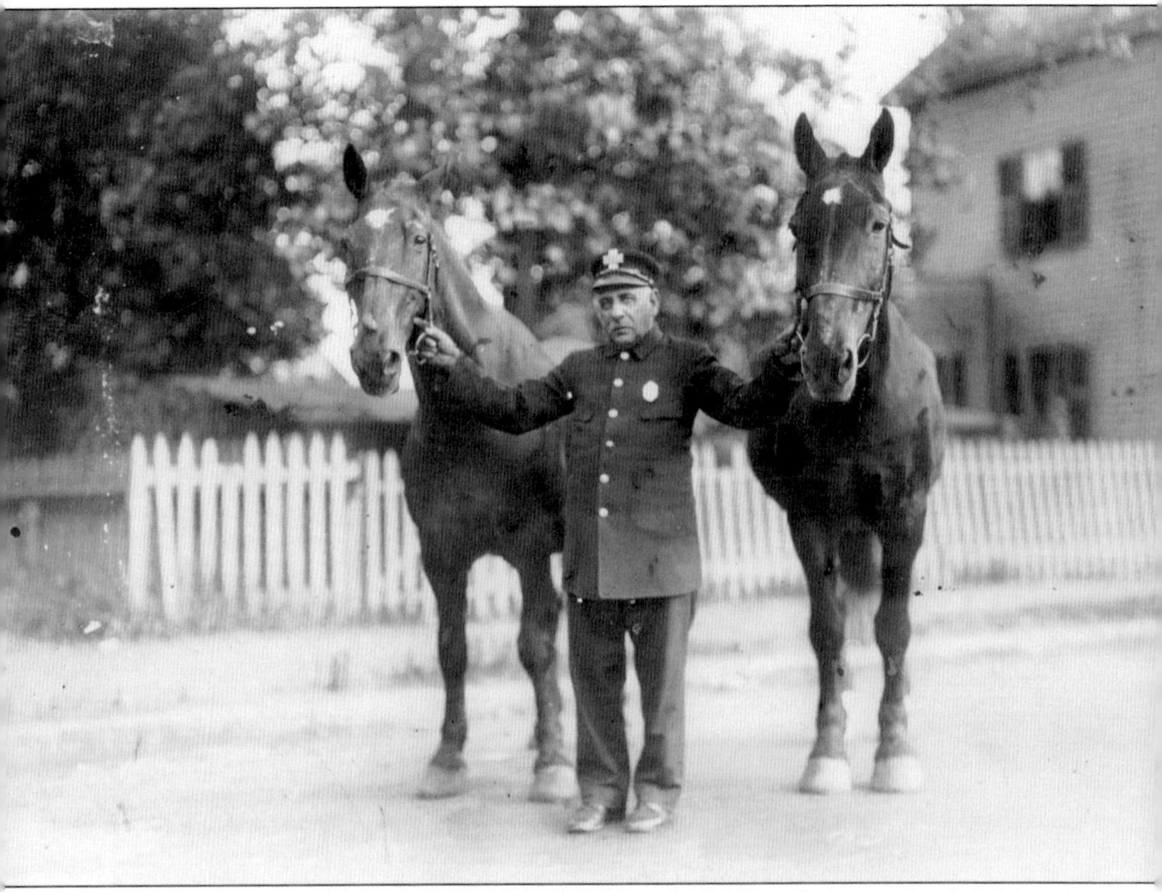

Horses played a key role in 19th- and early-20th-century firefighting. On September 2, 1926, engine No. 3 received its first motorized engine, thus bringing an end to the era of the horse-drawn fire apparatus. Featured here are Tom and Jerry with their driver and companion, Freeman Bodge, at their side before being retired with the knowledge and pride that they were once fire horses. They were the last two horses to be used by the Peabody Fire Department and were retired from service in 1926 by resolution of the city council.

Four

The Motorization Era

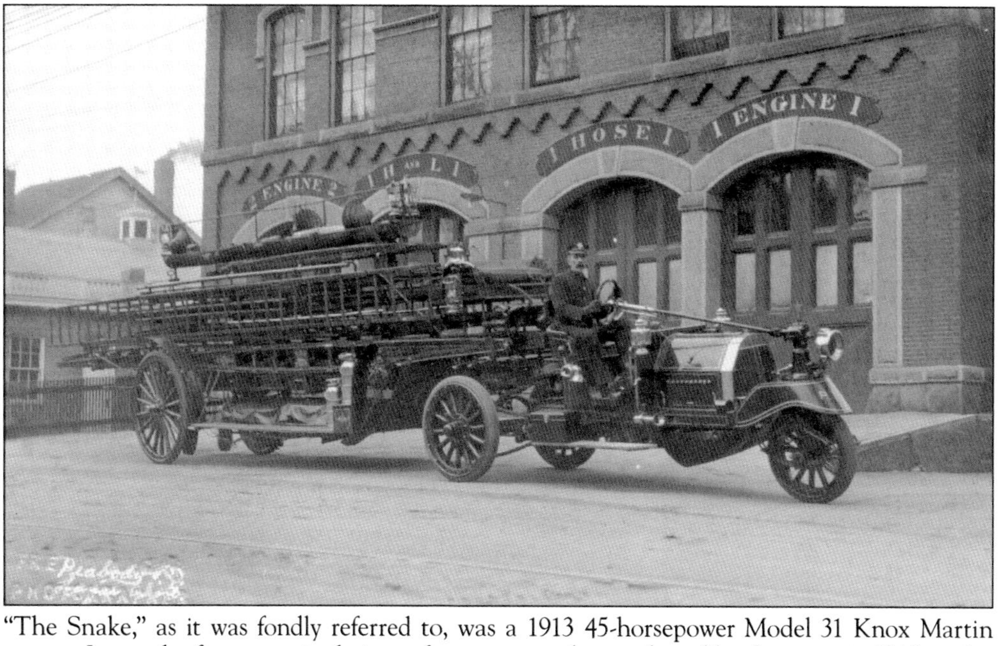

"The Snake," as it was fondly referred to, was a 1913 45-horsepower Model 31 Knox Martin tractor. It was the first motorized piece of apparatus to be purchased by the town in 1913, and it had three wheels. The tractor is attached to the 1908 horse-drawn ladder truck.

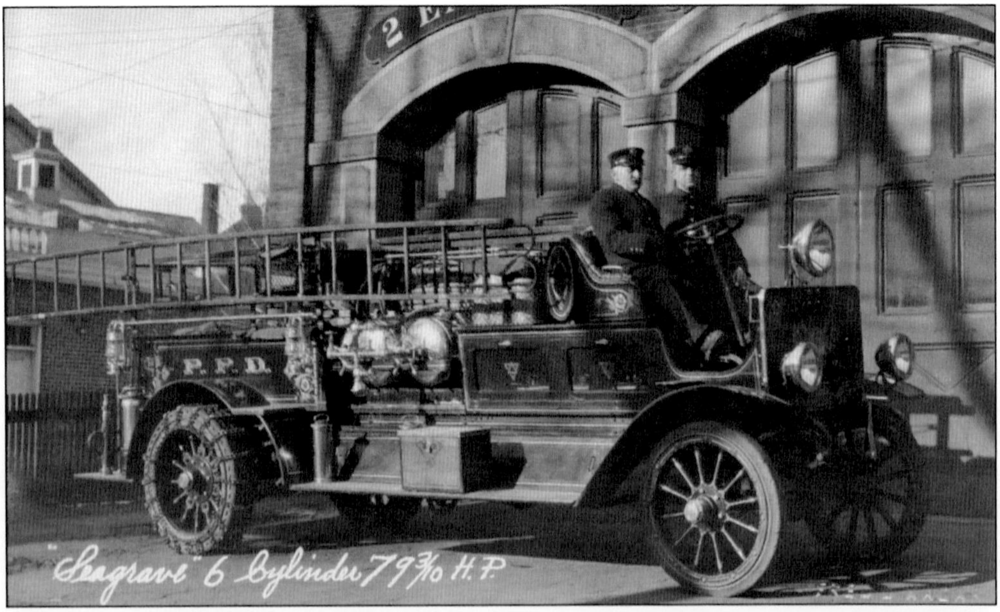

The second motorized piece of apparatus and the first motorized engine was also bought by Peabody in 1913. It was a six-cylinder, 79-horsepower automobile built by the Seagrave Company of Columbus, Ohio, at a cost of $6,000. It was known as a buckboard type, and it had ladders, lanterns, hoses, and two 36-gallon chemical tanks for putting the fires out. The department was not yet on the permanent two-platoon system, but some of the stations had permanent drivers. Pictured here are driver George Morgan (left) and Daniel Hannon.

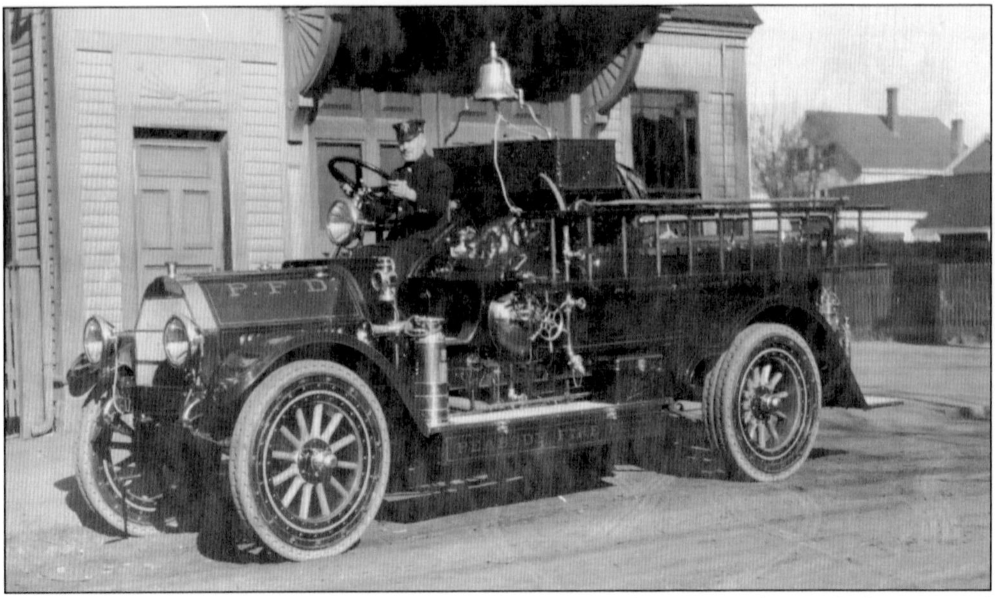

The firemen of engine No. 2 rallied to get their own motorized engine, and in 1915, they received this Knox model fire truck.

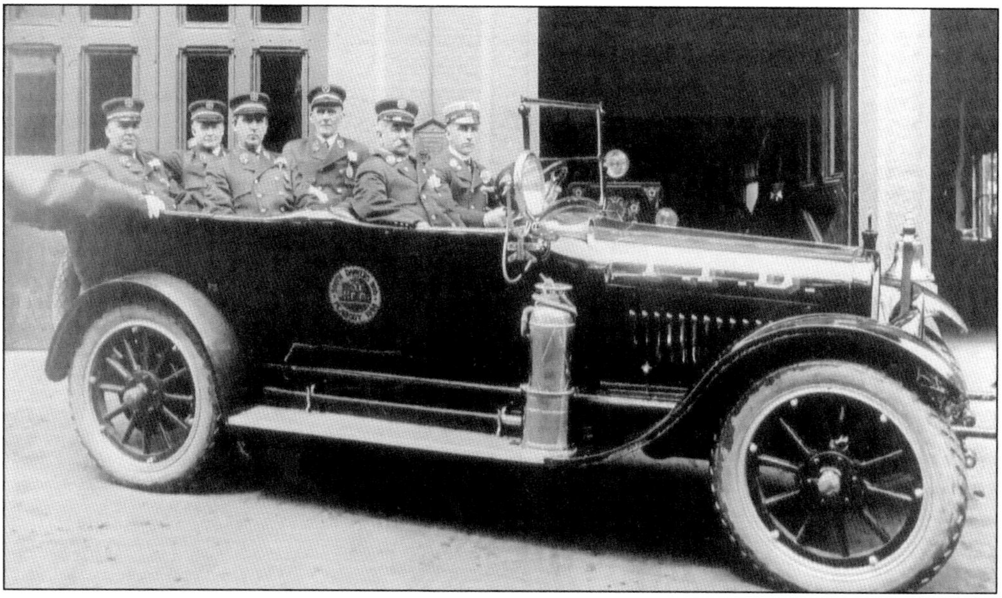

This picture from about May 1916 of the chief's new Chandler car shows Vincent P. O'Keefe, the superintendent of fire alarm and town electrician, driving chief engineer Jesse F. Barrett. Pictured in the rear seats, from left to right, are assistant engineers Arthur P. Bodge, Thomas Hutchinson, John Costello, and Thomas Carbrey. The car is a Chandler, from the Chandler Motor Car Company of Cleveland, Ohio. W. J. Fahey, of the Peabody Garage on Hardy Street, was awarded the contract of $1,250 for this seven-seat car that was used for the chief and fire alarm superintendent. This type of car was decided upon because it was capable of carrying firemen when necessary. The car was painted red, was labeled with the town seal, had a locomotive bell and siren horn, and was equipped with a hook, axes, and chemical tanks.

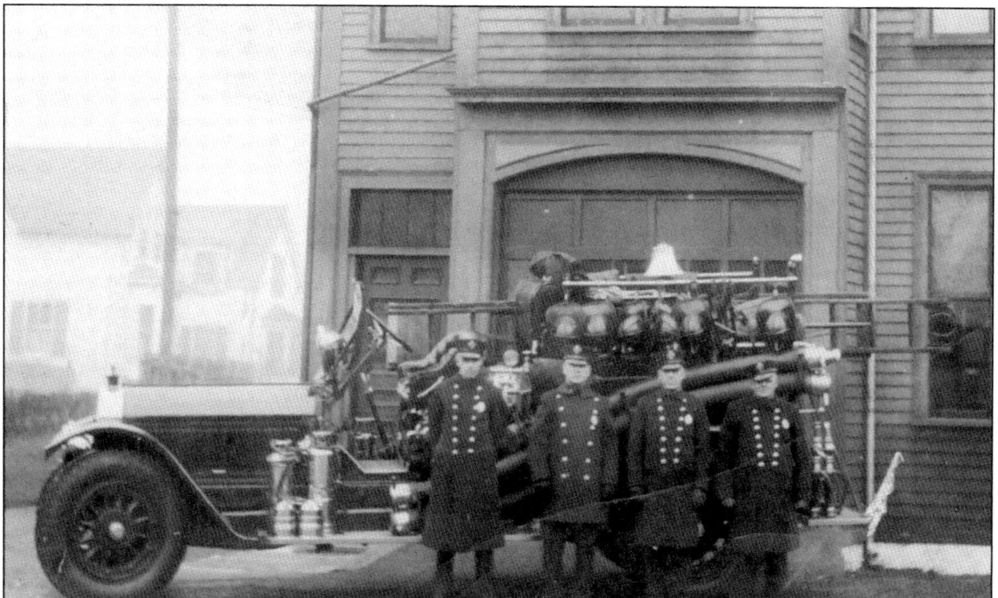

Engine No. 5 received their first motorized engine in 1916. It was the first triple combination engine that the city purchased.

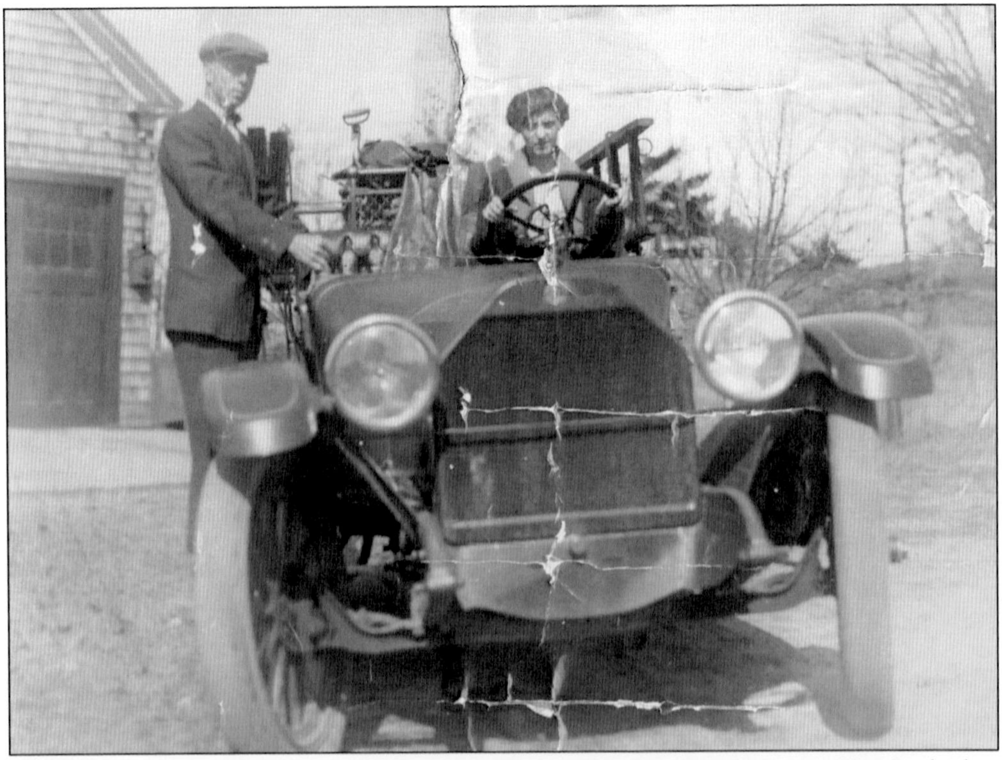

Seen here are Ruby and E. Lawrence Durkee in engine No. 7 on Pine Street in West Peabody.

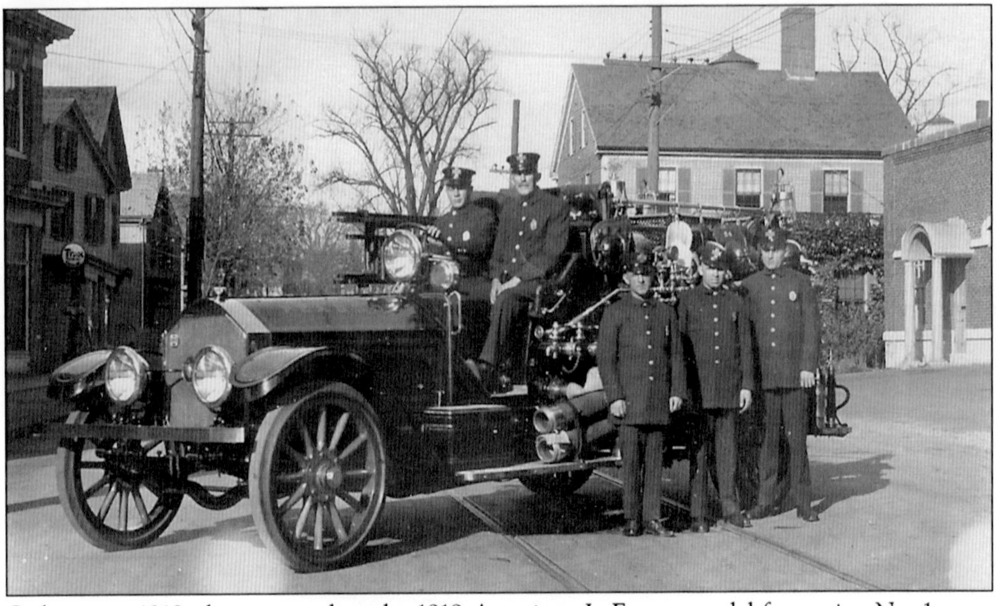

Ordering in 1918, the city purchased a 1919 American LaFrance model for engine No. 1, seen here with its crew on Lowell Street in front of Central Station. It was a triple combination engine that arrived on April 6, 1919. It was red and gold with the number one painted on it. Pictured from left to right are Ray Skinner, William Brown, Fred Sherry, Charles Lawrence, and Arthur C. Flynn.

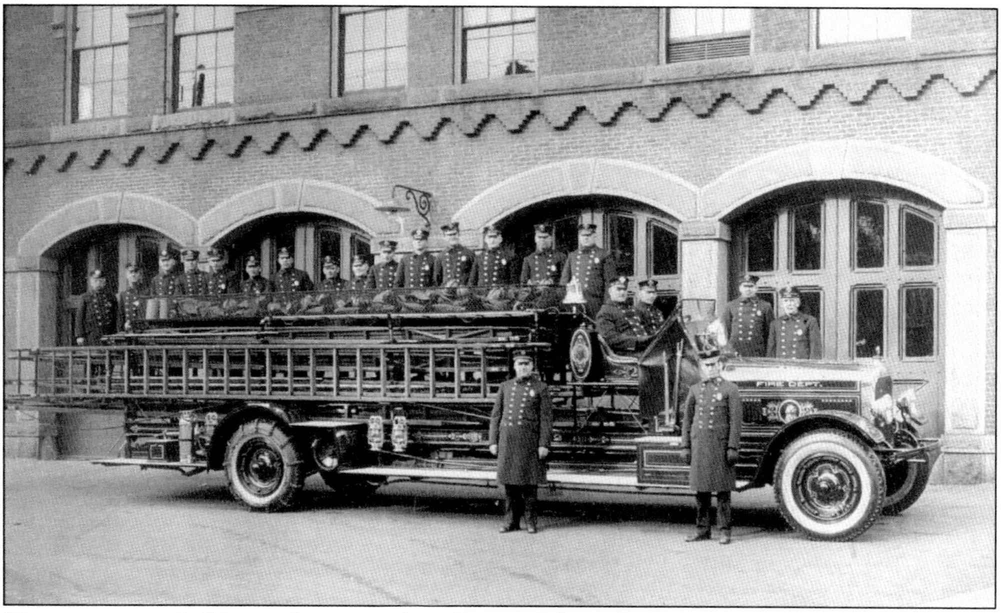

This city service ladder truck was purchased in 1923 from the Seagrave Company of Columbus, Ohio. This truck serviced the town well until the purchase of an aerial ladder truck in 1946. Aerial ladder trucks had extendable ladders attached to them by a platform that was raised hydraulically. Ground ladders were placed in racks underneath the aerial ladder. Today's ladder trucks are bigger than ever, and some even carry water like an engine company does. They are called quints because they have five components, including a pumping engine, a booster tank, hose, ground ladders, and an aerial ladder.

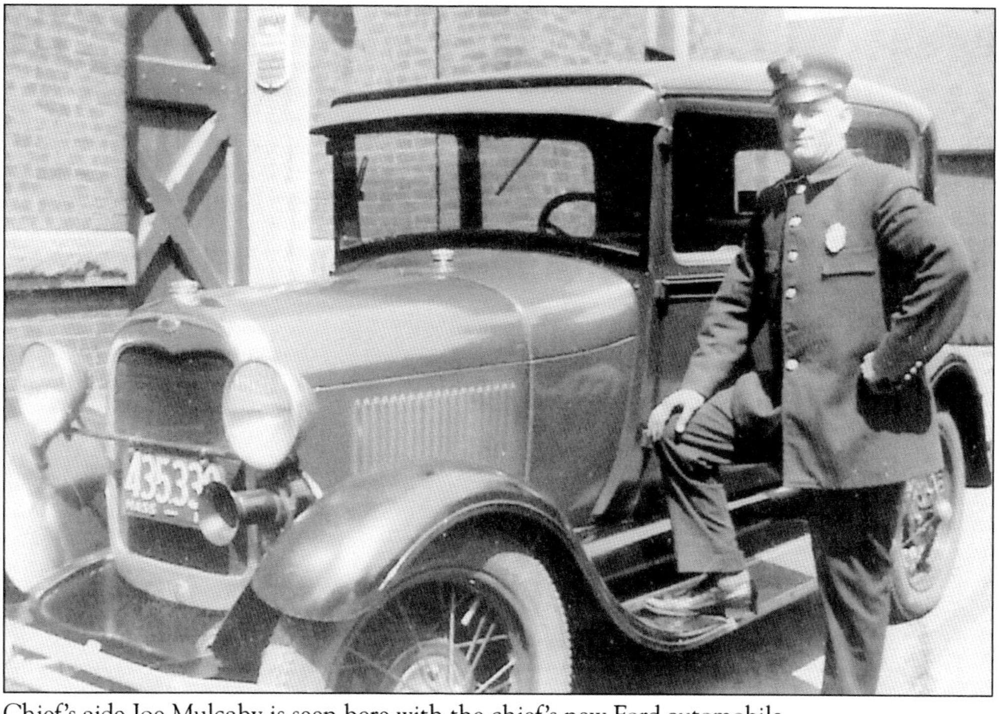

Chief's aide Joe Mulcahy is seen here with the chief's new Ford automobile.

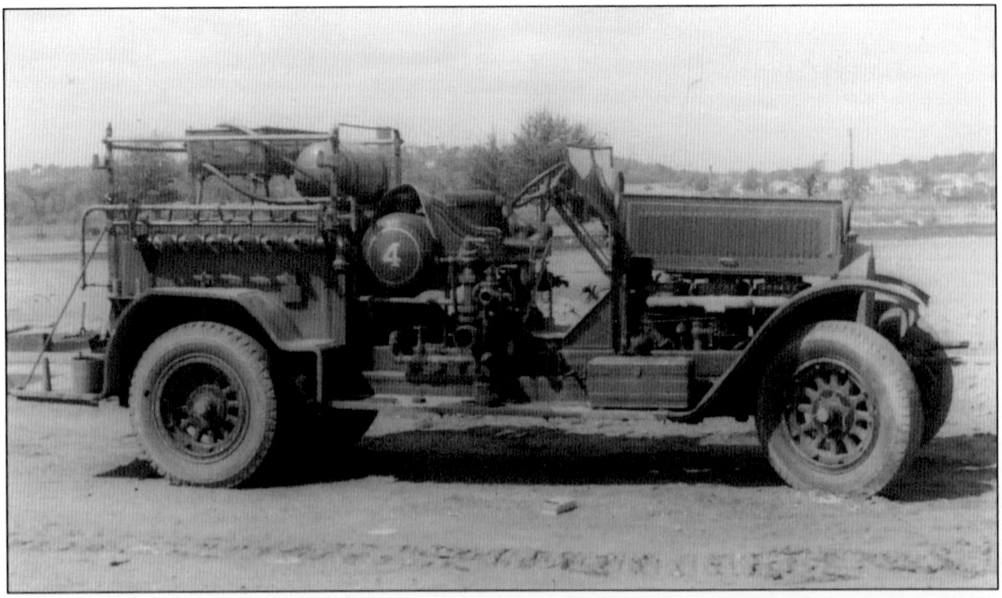

Engine No. 4, a 1925 chain drive pump, was built by the American LaFrance Company. It was photographed in Saugus at the dump opposite Bristow Street on September 18, 1953. It arrived on August 17, 1925, and cost $12,250. It was a 750-gallon triple combination pumper with a 60-gallon booster tank. Among the many tools that engine No. 4 carried in its compartments, one in particular was the P and Q bar. This forcible entry tool was invented by two Peabody men. (Courtesy of the Peabody Historical Society; photograph by Laurence I. Beake.)

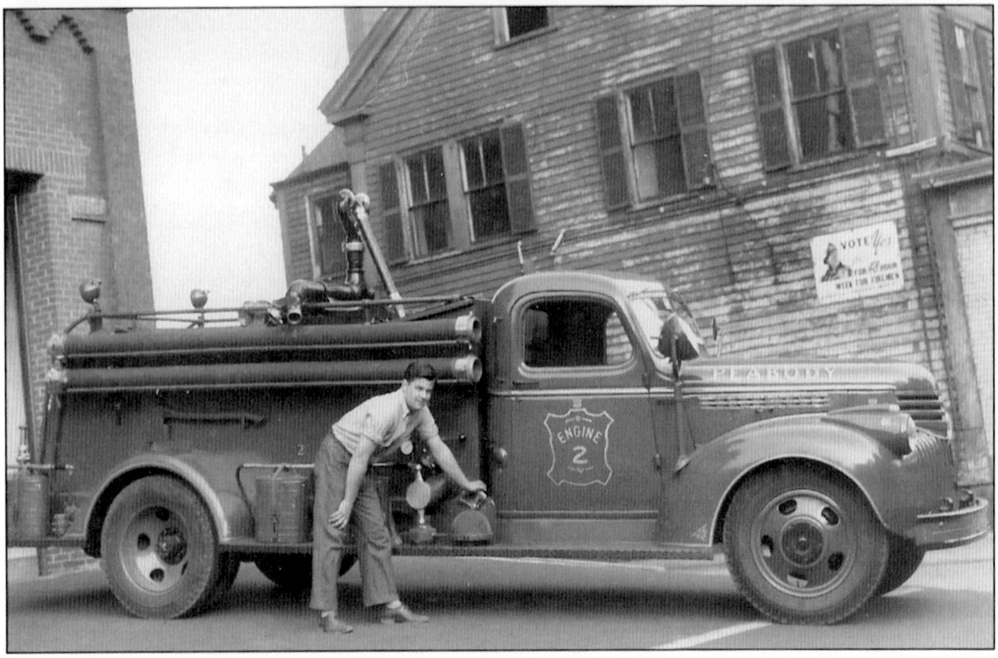

Ray Nelson, one of the few Peabody Fire Department mechanics, poses with a 1942 World War II U.S. Army surplus Chevrolet engine. It had a 375-gallon-per-minute pump and was later renumbered engine No. 2 in 1962 and engine No. 6 in 1965. It was then used as a lighting truck in the early 1970s before being sold to Art Blake around 1971.

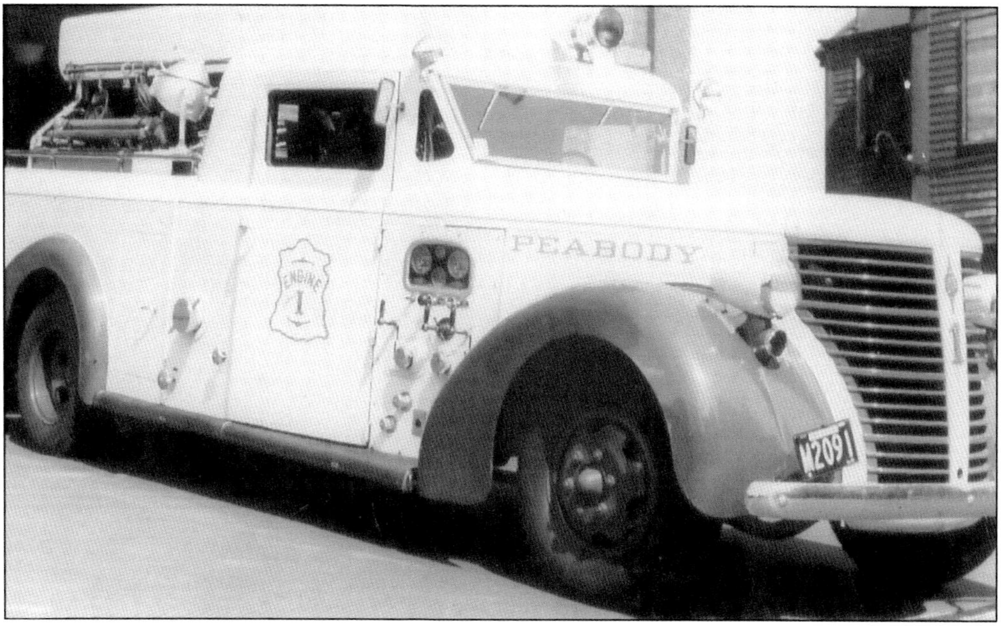

In 1940, the city purchased two new engines. The chief and the city's purchasing agent were at odds on what manufacturer was better, so they finally compromised. Engine No. 1 was an all-white 1940 American LaFrance Series 500, with a 750-gallon-per-minute pump. It was placed out of commission in 1949 when it had an accident in Peabody Square while responding to an alarm on Lynn Street. Engine No. 1 had many nicknames, such as "the Milk Wagon," "the White Ghost," and "the White Angel." It was also referred to as "the White Elephant."

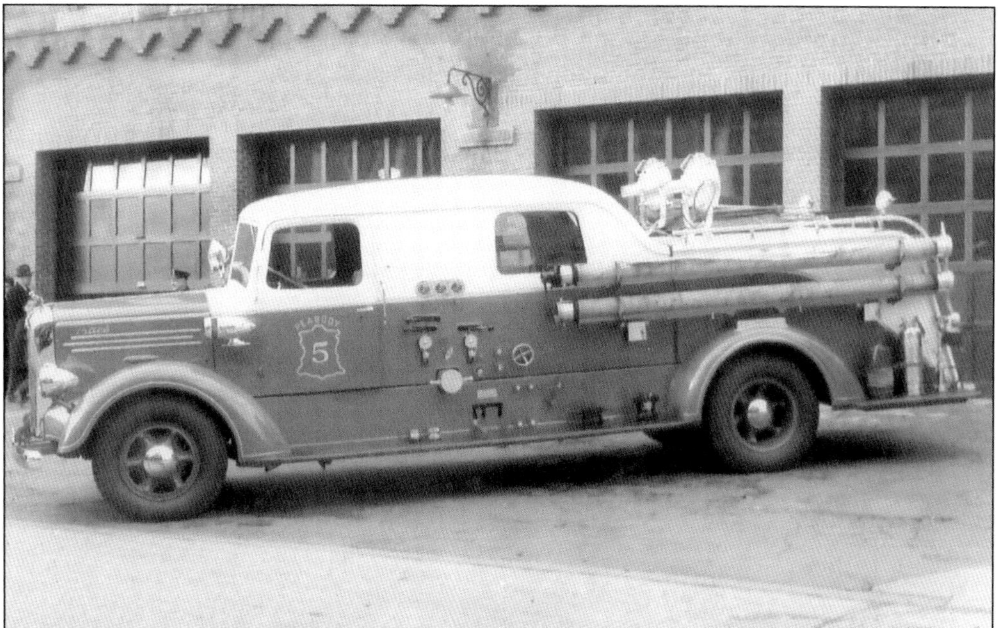

Engine No. 5 was a 1940 Mack Model 85L chassis with a 750-gallon-per-minute pump, a red body, and a white roof. It is seen here in front of headquarters. It was destroyed in the April 14, 1961, fire at the engine No. 5 firehouse on Lynn Street.

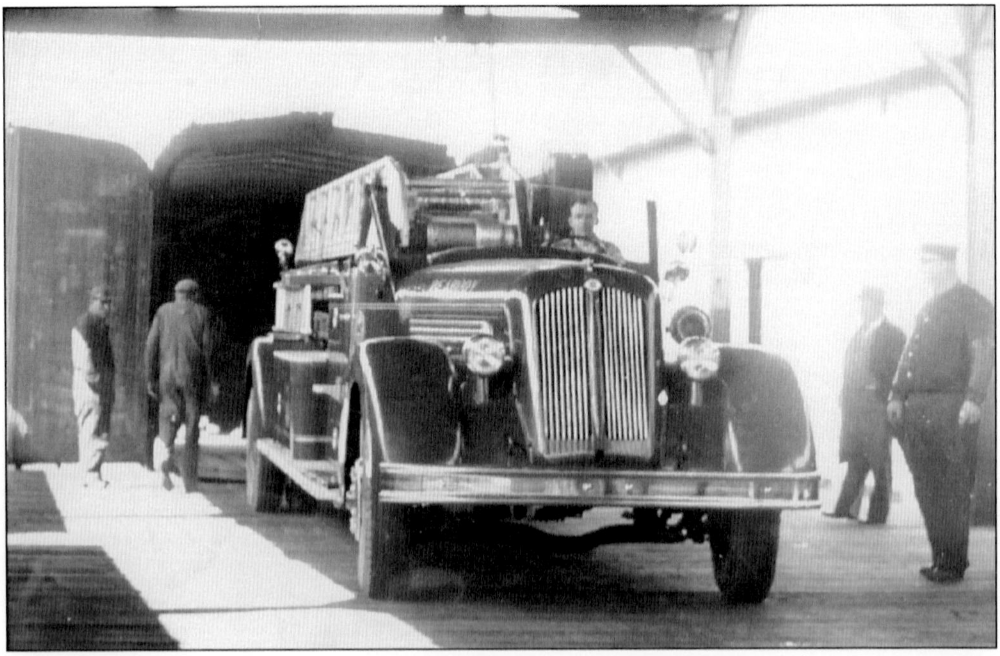

Peabody's first aerial ladder truck was purchased in August 1946 at a cost of $26,000. Its arrival was delayed many times until it finally happened in March 1949. Seen here is a very rare photograph of the 1946 Seagrave ladder truck, nicknamed "Agnes," as it is being driven out of a railroad boxcar somewhere in Boston.

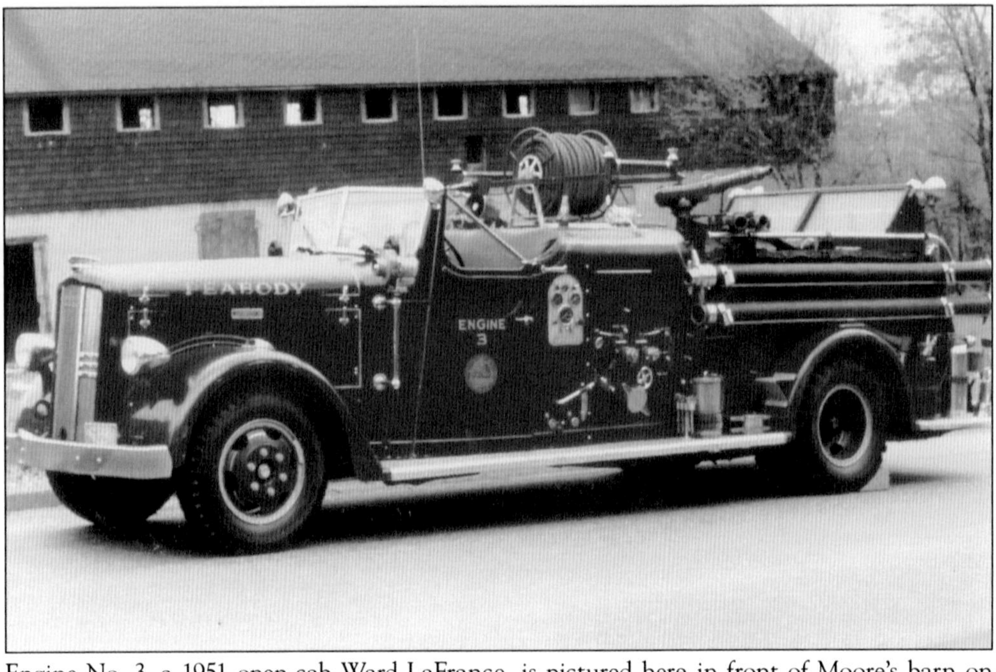

Engine No. 3, a 1951 open-cab Ward LaFrance, is pictured here in front of Moore's barn on Lowell Street in West Peabody.

for MAY, 1952 327

WARD LaFRANCE Pumpers

. . . for Peabody, Mass.

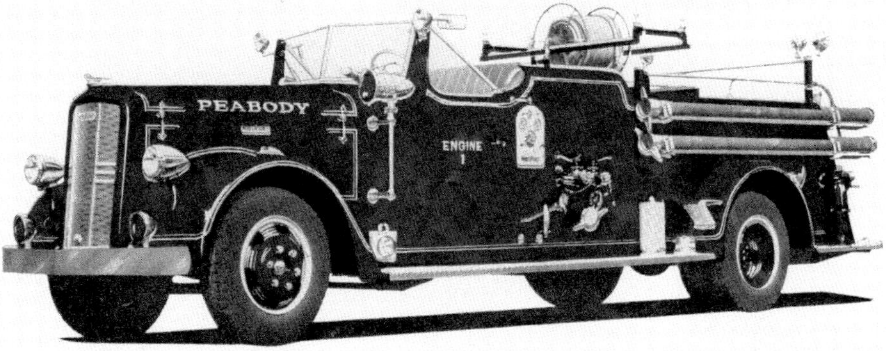

PEABODY, MASS. — a great industrial and residential city — is one of the fastest-growing communities in the Bay State thanks to its accessibility to main highways and railroads. Like so many cities, this New England community relies on Ward LaFrance for fire protection.

The reason is simple. Through the years, Ward LaFrance Pumpers have demonstrated their ability to perform under a wide range of conditions. These sturdy pumpers come with engines in 130, 150, 200 or 240 H.P. — deliver the weight-to-horsepower ratio so necessary for quick get-aways and climbing hills.

Every Ward LaFrance Pumper features a Waterous split housing pump — surplus gallonage that far exceeds rated capacity — flat horizontal booster tank — exceptional hose capacity — and other equally famous features.

FREE BULLETIN There's a new bulletin that gives full details on Ward LaFrance Fire Apparatus. Send for it today!

New England Agent: I. A. Ballou, Jr.
Dover & Allerton Avenue
East Providence 14, R. I.

WARD LaFRANCE

TRUCK CORPORATION

Fire Apparatus Manufacturers for over 30 Years

GENERAL OFFICE & FACTORY: Grand Central Avenue, Elmira, N.Y. *Tel. 5171*

DIRECT FACTORY BRANCH: 139th Street & Bruckner Blvd., New York 54, N.Y. *Tel. CYpress 2-3300*

EXPORT: Ward LaFrance International, Inc., Elmira, N.Y., U.S.A. *Cable Address: "WARDTRUCK"*

In 1951, Peabody purchased three Ward LaFrance fire engines. These 750-gallon-per-minute, open-cab fire engines were assigned to engine Nos. 1, 3, and 4. Two of them arrived on March 24, 1952. The third one arrived sometime soon after. Engine No. 1 became engine No. 2 in 1965 and then sold in 1975. Engine No. 3 became engine No. 6 in 1971 and then sold in 1982. Engine No. 4 was also sold in 1975. This advertisement was in one of the firemen journals in 1951.

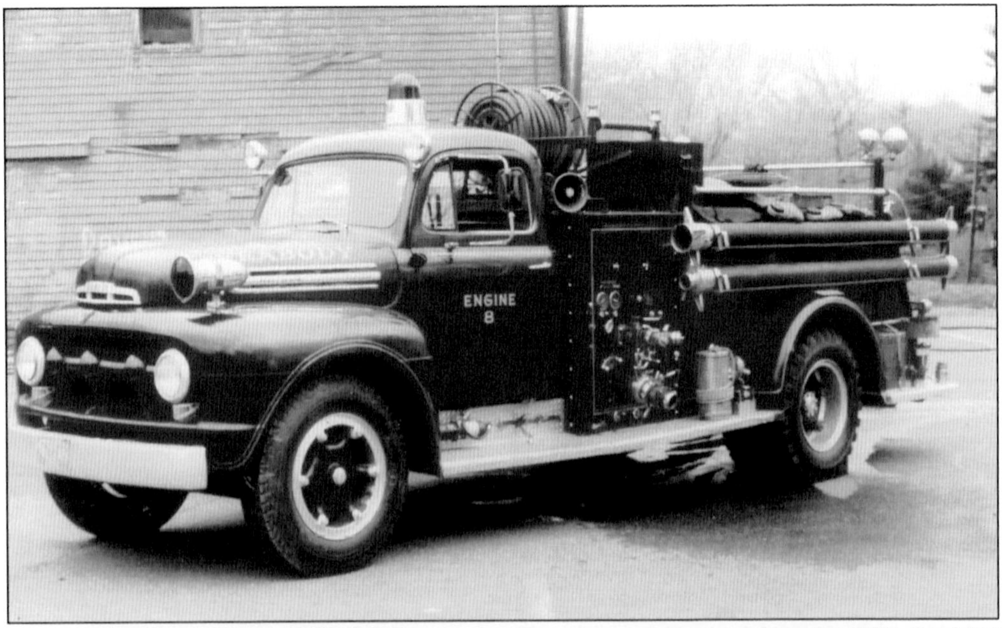

Along with the three new Ward LaFrance engines, the city also purchased a 1951 Ford F-7 Ward LaFrance engine. It was a 500-gallon-per-minute, closed-cab forestry truck for engine No. 8. It was sold in 1982 to Gerald Nelson of Madison, Maine. Arthur Green, who was the last forest warden, drove it. The position of forest warden was abolished, and the responsibilities fell entirely upon the chief of the department.

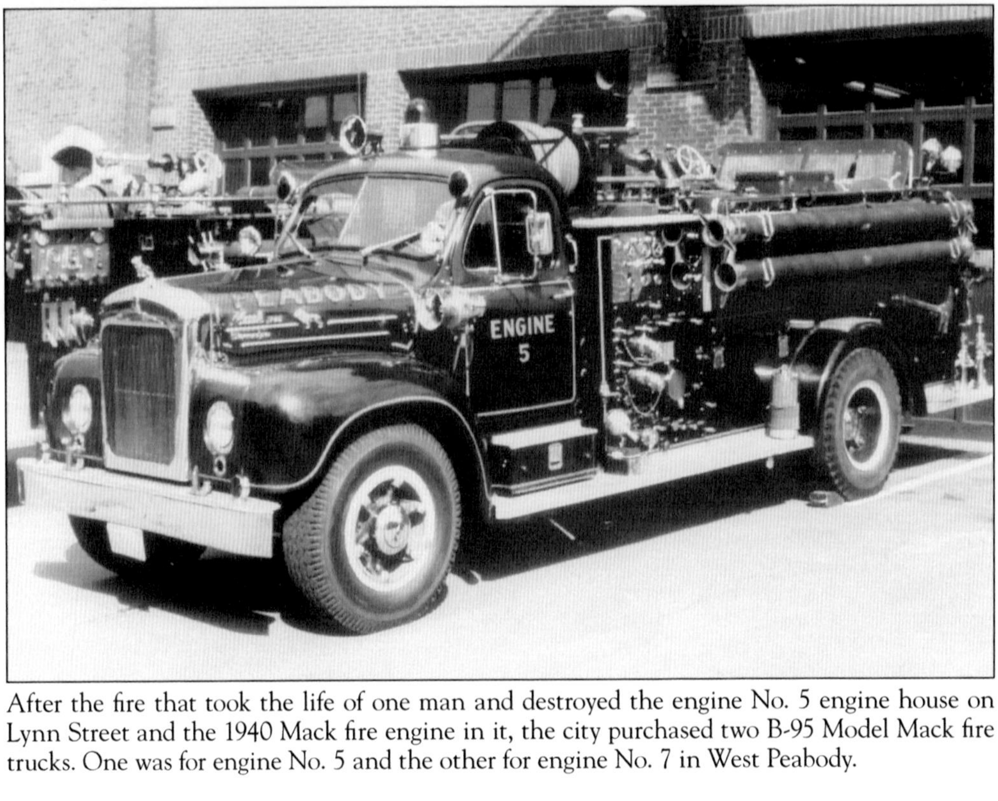

After the fire that took the life of one man and destroyed the engine No. 5 engine house on Lynn Street and the 1940 Mack fire engine in it, the city purchased two B-95 Model Mack fire trucks. One was for engine No. 5 and the other for engine No. 7 in West Peabody.

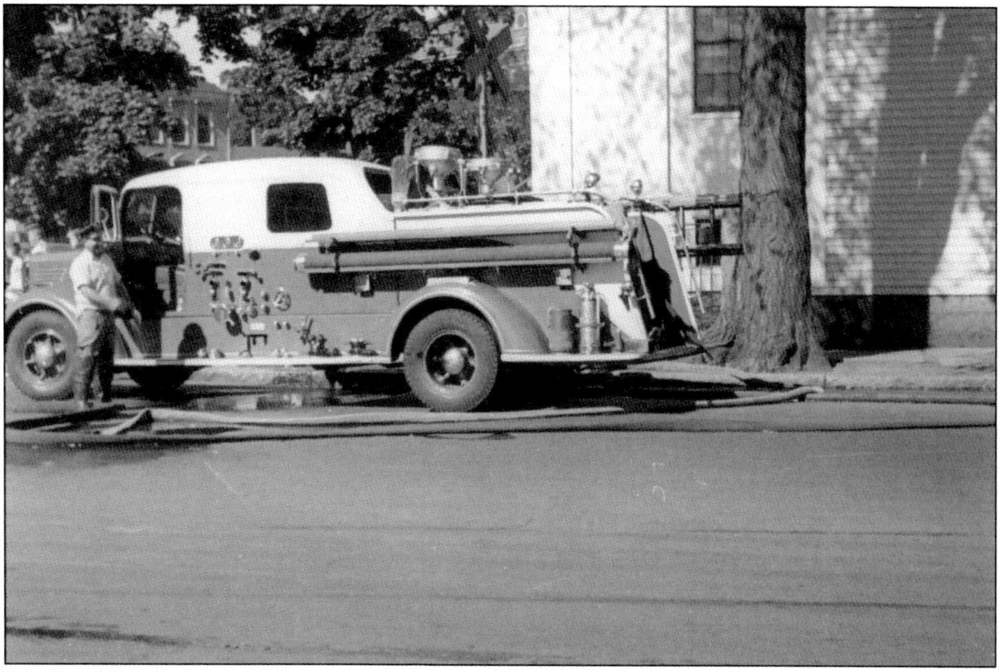

Engine No. 5 is seen here pumping a fire in Peabody Square next to the South Congregational Church.

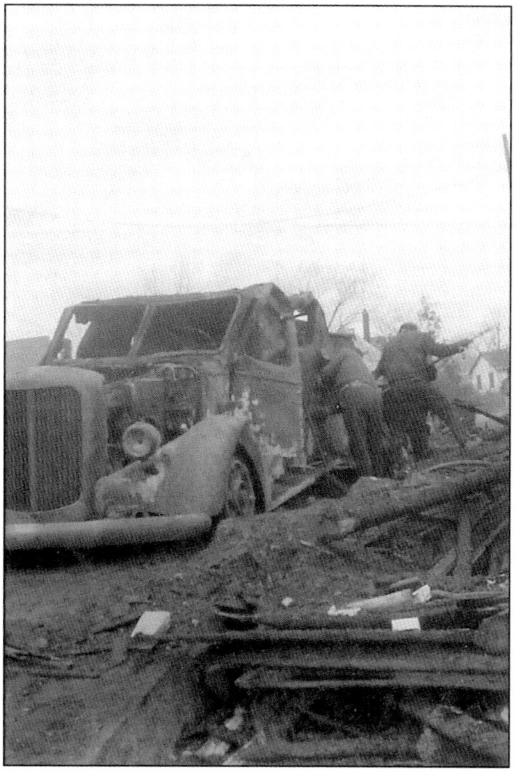

It was not the first station in Peabody to have a fire, but it was the only one that was a total loss. Engine No. 5 on Lynn Street burned to the ground on April 14, 1961. The only surviving piece of the building was the fire bell, seen on page 89 crashing into the building. It was in tough shape with melted tar, shingles, and a big crack, but after years of storage, the bell was finally restored and sits out in front of the station as a tribute to the courage and tenacity of the firefighters of Peabody.

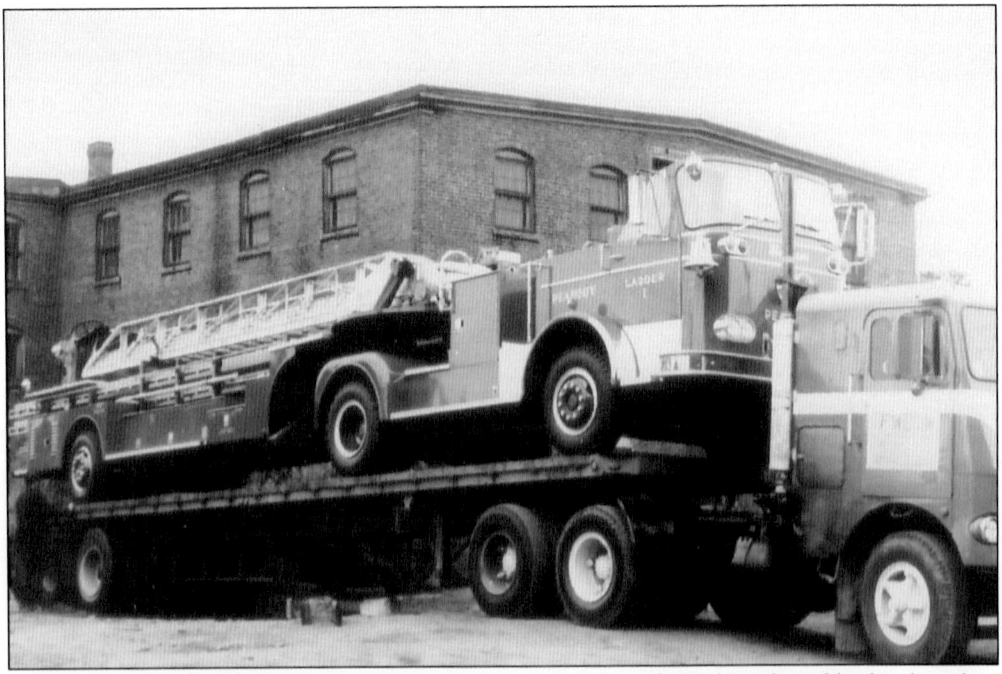

In 1967, the city purchased a new ladder truck. It arrived on a flatbed truck and had to be taken off at the freight terminal building on Foster Street. It was a 1967 Seagrave Model 800 MT. It was an open-cab ladder truck with an 85-foot aerial ladder.

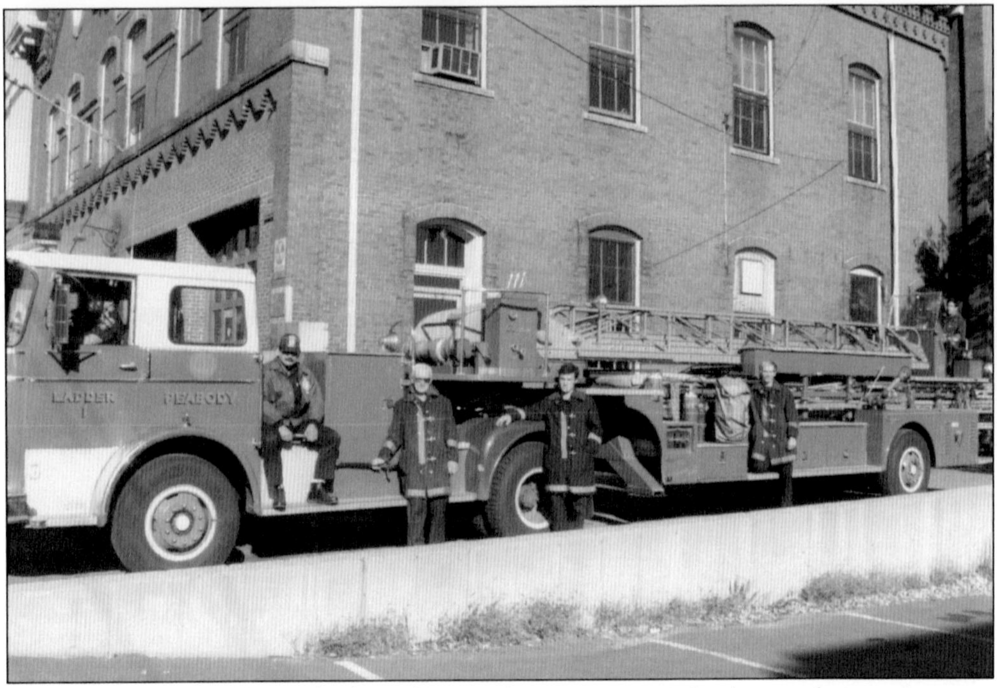

The 1967 Seagrave ladder truck, shown here, underwent some overhauling, and a cap was put on the cab to enclose it. Pictured from left to right are unidentified, Rick Curris, Jim Kerwin, Paul Goglia, John Kerwin, and Michael Zuppio driving the tiller.

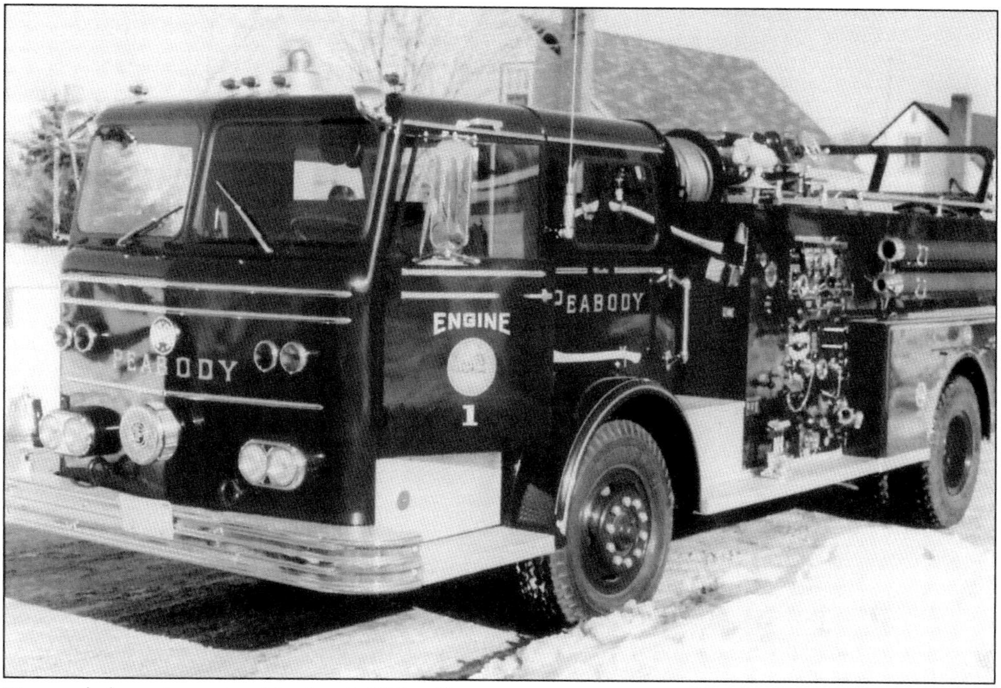

Pictured above is engine No. 1, a 1,000-gallon-per-minute pumper. The total price for this piece, with the trade-in of the 1965 Seagrave, was $42,000, a far cry from the first motorized purchase of $6,000 made 50 years prior. Pictured below is ladder No. 2. The city purchased these two pieces of apparatus, engine No. 1 and ladder No. 2, at the same time in 1970. The engine arrived in about six months on December 4, 1970; however, the ladder truck did not arrive for about two years. It showed up one night at the engine No. 7 firehouse. It was an open cab, and the driver had driven it, in the winter, from Wisconsin. He made some kind of wooden box contraption to protect himself from the elements. The reason it took so long was that the chief wanted a mid-mount ladder only. When it did arrive, it was a mid-mount ladder mounted on a rear-mount chassis. This caused the trailer to always tilt down toward the tractor.

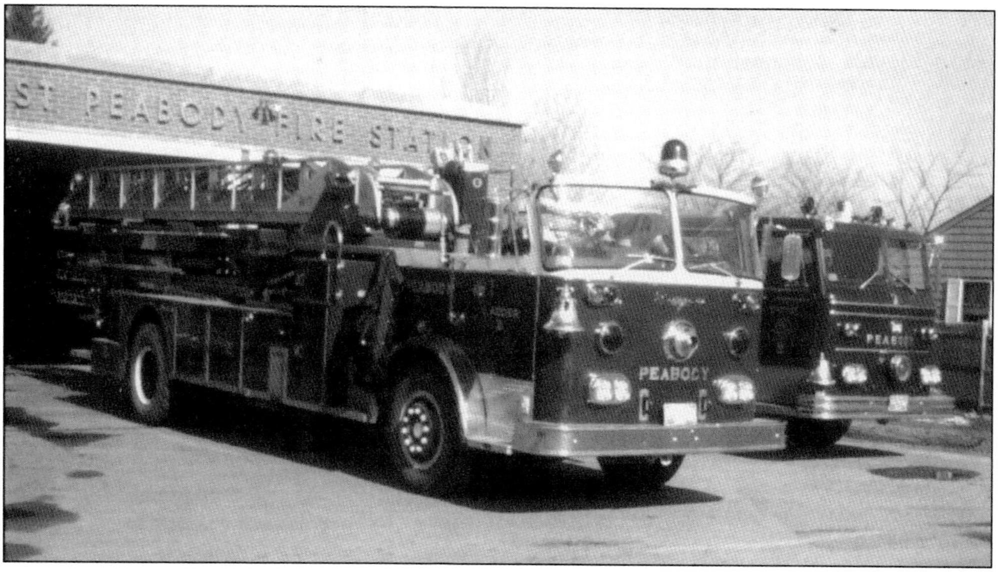

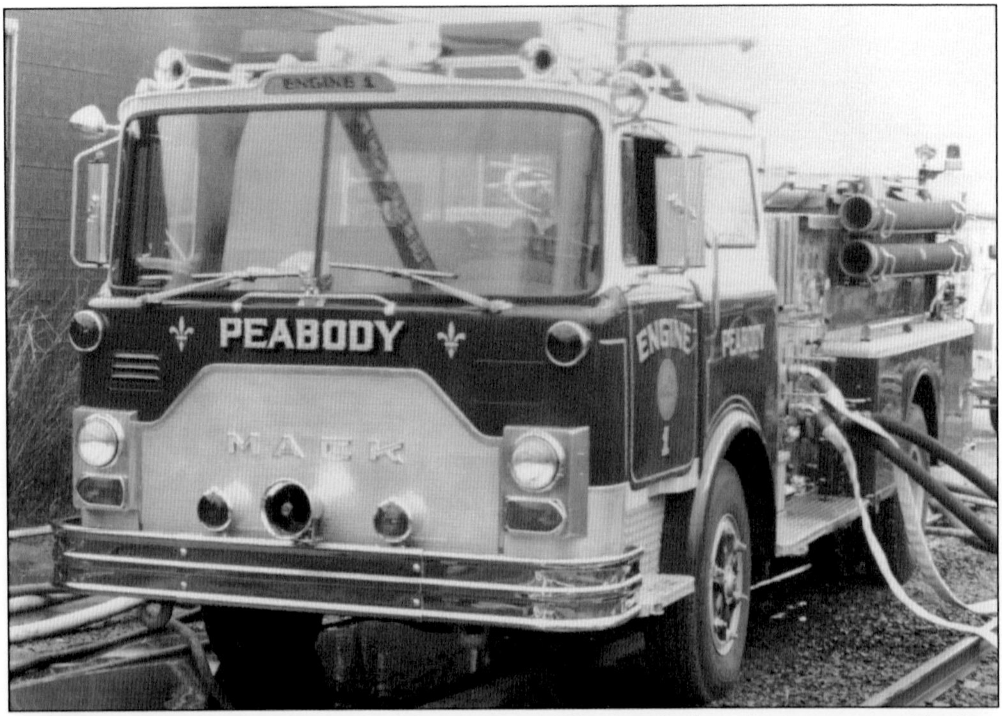

A 1986 CF600 Mack became the new engine No. 1. This engine later became engine No. 3 in the 1990s and then engine No. 5. It is now engine No. 2, a reserve engine.

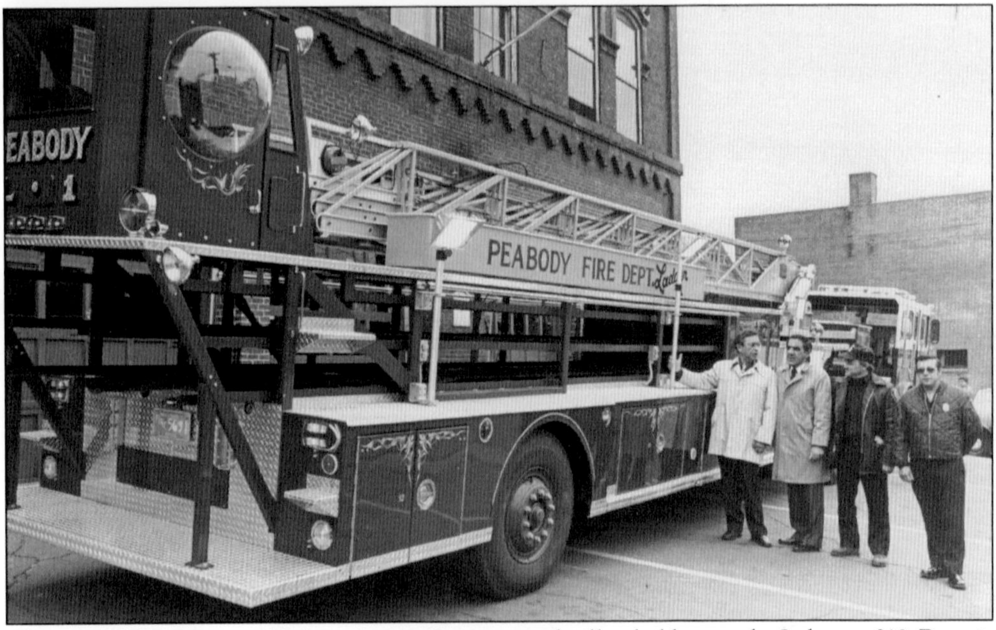

This 1986 Maxim is a 100-foot tractor-drawn aerial tiller ladder truck. It has a 318 Detroit Diesel and an automatic Allison transmission. It is now ladder No. 3 stationed at the engine No. 3 engine house at 5 Prospect Street. It is now only 90 feet and is a reserve ladder. Pictured here from left to right are Chief Nicholas Gerakaris, Mayor Peter Torigian, Patrick Labbe, and firefighter William "Beaver" Reynolds.

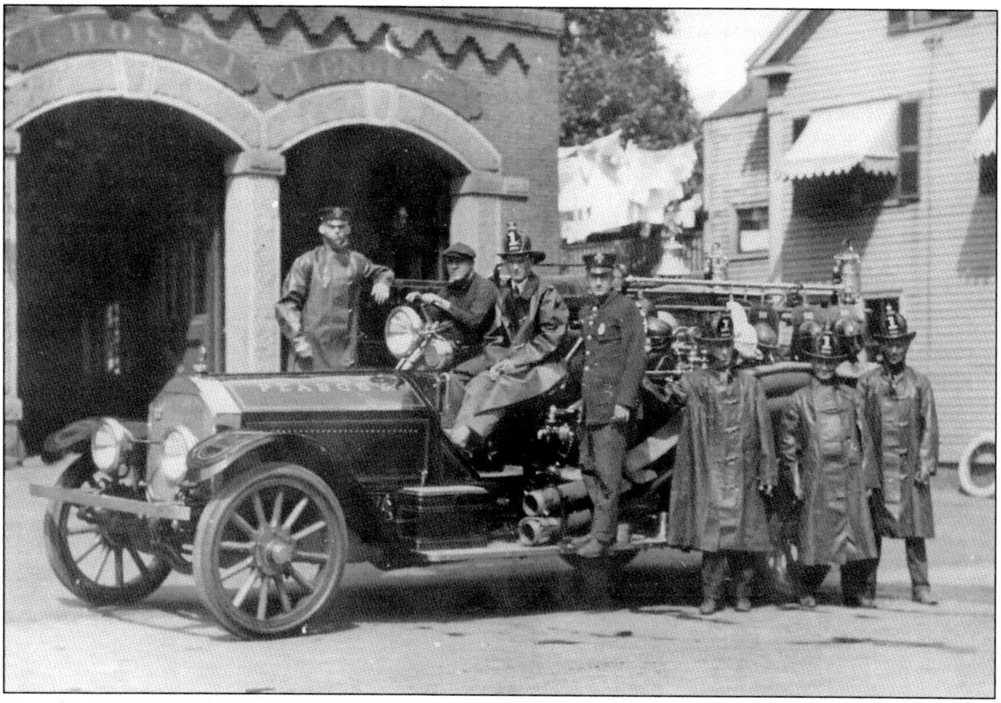

Seen here are, from left to right, J. Roche, J. Carr, J. Mularky, C. Duggan, W. Mahoney, C. Caden, and R. Smith.

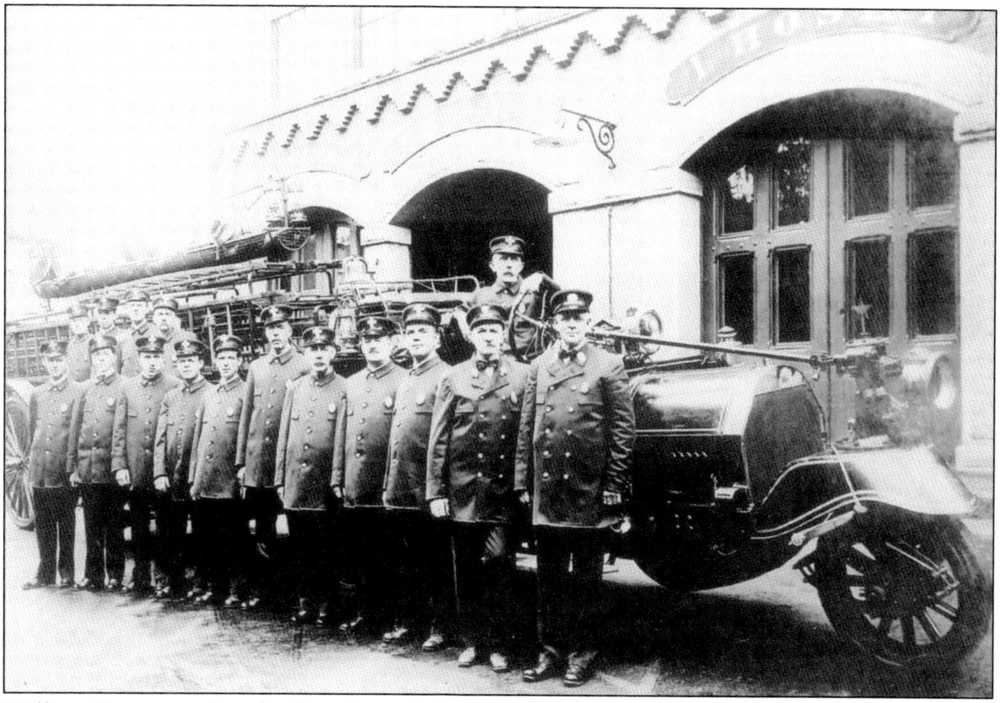

William Brown, at the wheel, and members of the ladder company are pictured here in front of Central Station. This ladder truck was nicknamed "the Snake."

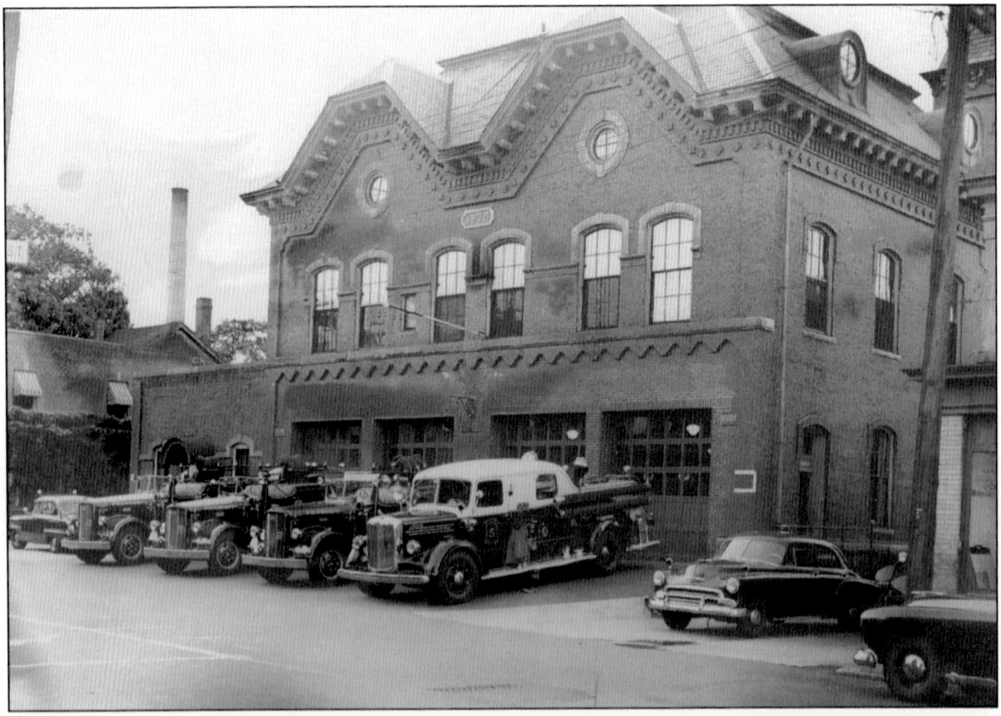

Pictured in front of Central Station are the three 1951 Ward LaFrance engines, engine Nos. 1, 3, and 4, and the 1940 engine No. 5 Mack engine.

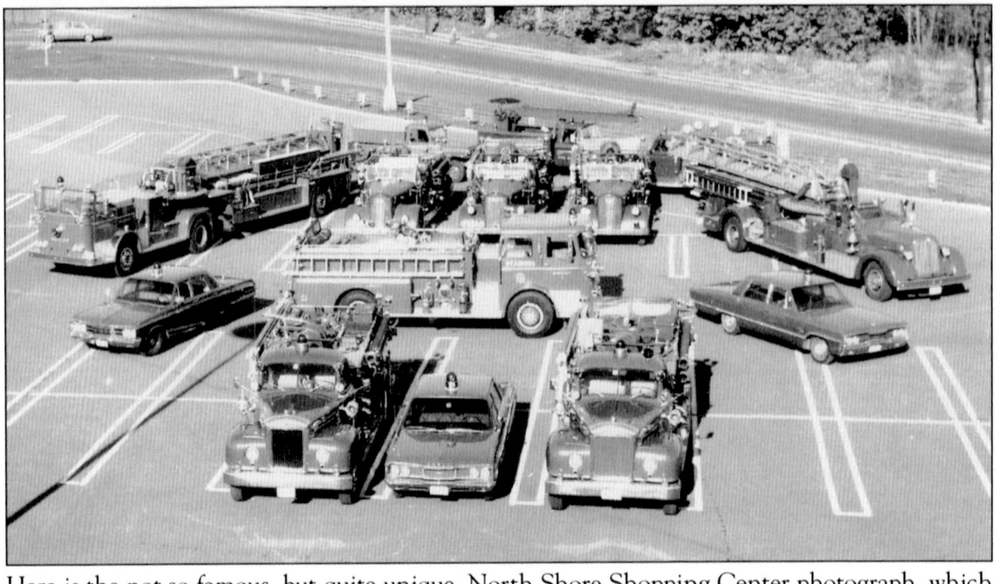

Here is the not-so-famous, but quite unique, North Shore Shopping Center photograph, which was taken sometime in 1972. This photograph was taken at the shopping center, and it had all the pieces of apparatus that belonged to the Peabody Fire Department. It got outside cities to cover the stations while it took this picture. From left to right are (first row) engine No. 7, the chief's car, and engine No. 5; (second row) the deputy's car, engine No. 1, and the fire prevention car; (third row) ladder No. 1, engine Nos. 3, 4, and 6, and ladder No. 2; (fourth row) the mechanic's Jeep Whilly, the fire alarm truck, and a Chevrolet chassis lighting truck.

Five

Fire! Fire!

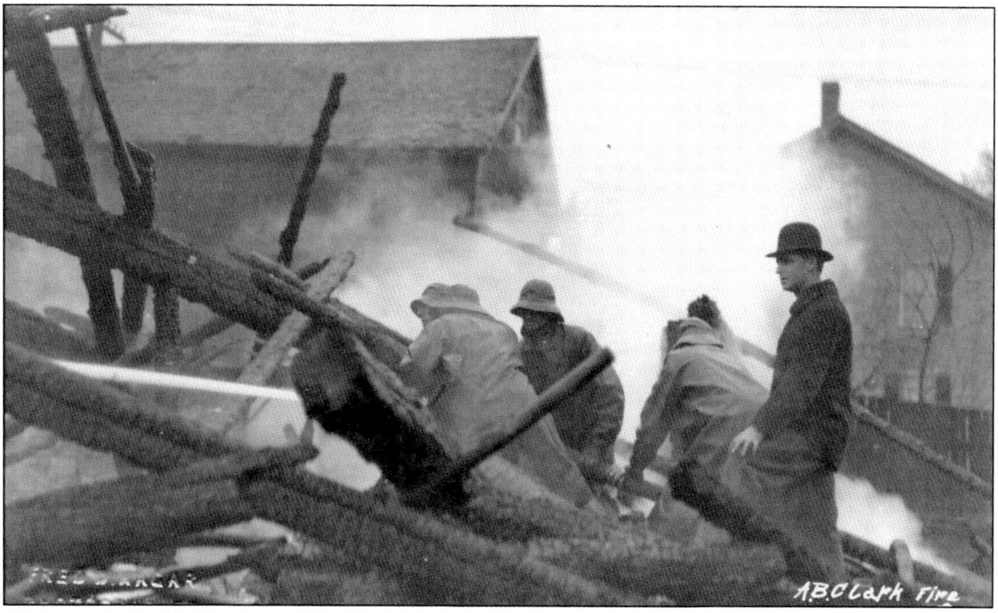

On a rainy April 28, 1909, a fire broke out at the A. B. Clark sheepskin factory. Although the rain helped stop the further spread of the flames, the damage caused by the fire was the largest to date with the loss valued at over $300,000. Photographs of this fire are, quite possibly, the first pictures of a fire in Peabody.

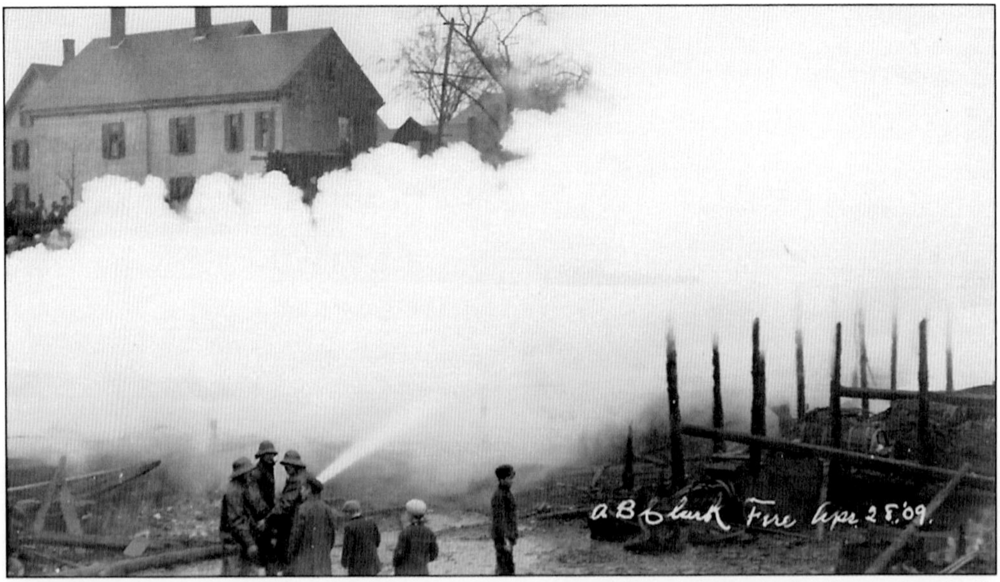

The photograph above shows the men of Hose Company No. 1 fighting the fire at the A. B. Clark sheepskin factory. Prior to the invention of motorized fire engines with pumps, the hose companies laid hose lines directly to the hydrant and relied on the hydrant's pressure. Note the firefighting gear, or turnout gear as it is referred to today. They wore rubber coats and leather or rubber hats. The photograph below was taken the day after the fire and shows the extent of the damage caused by the fire. The factory was located off Walnut Street in the east end of town and was nestled in between English and Union Streets.

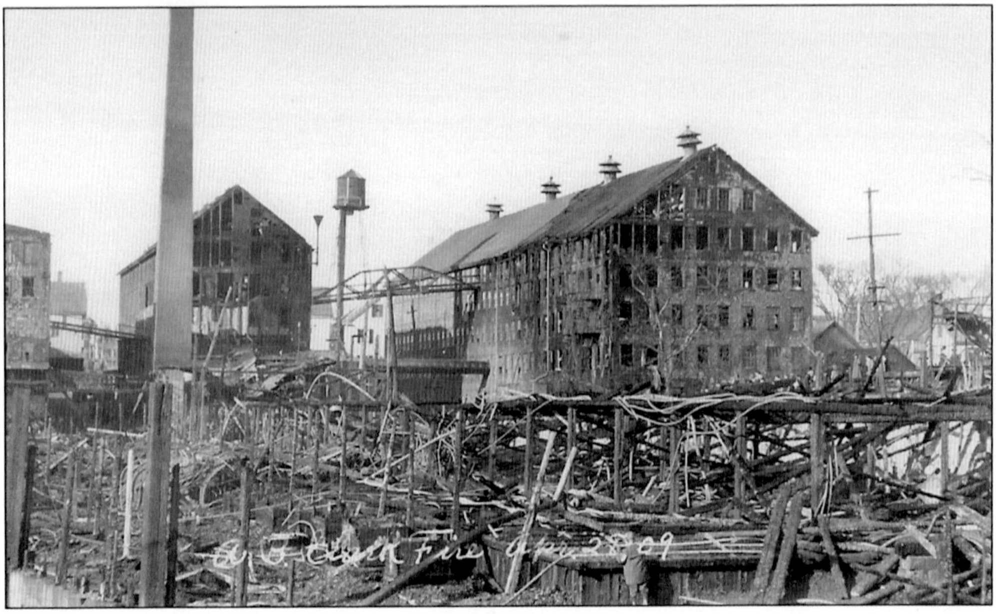

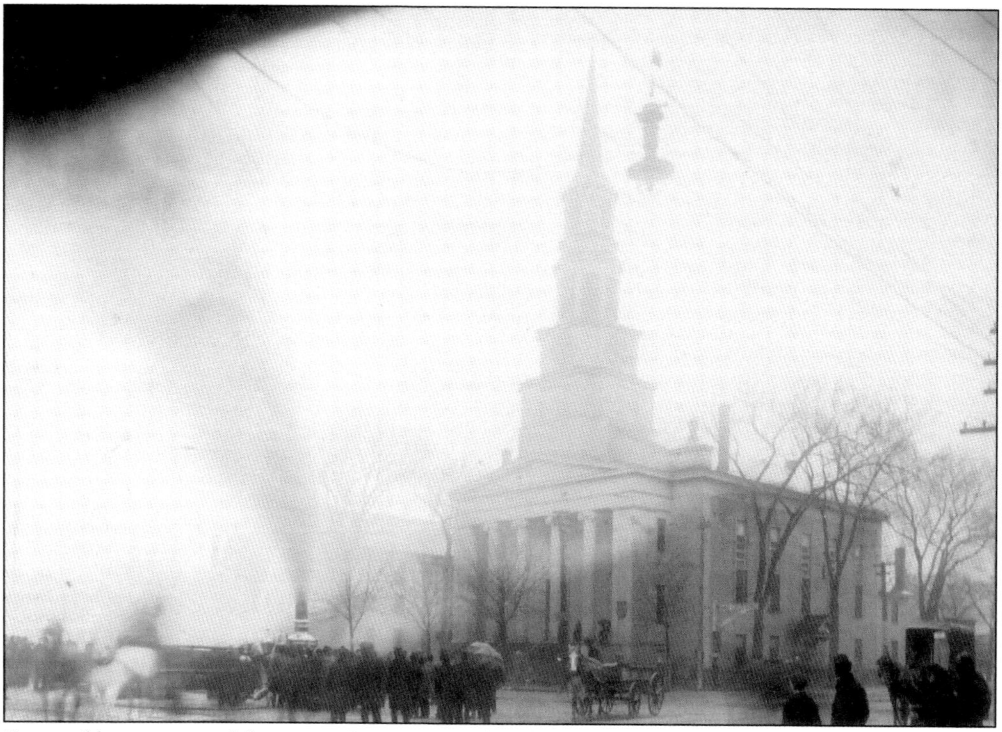

Pictured here is one of the steam fire engines of Peabody, with its black billowing smoke, working in the square. It appears that there is a fire next to the South Church in the A. H. Whidden's Hardware store building.

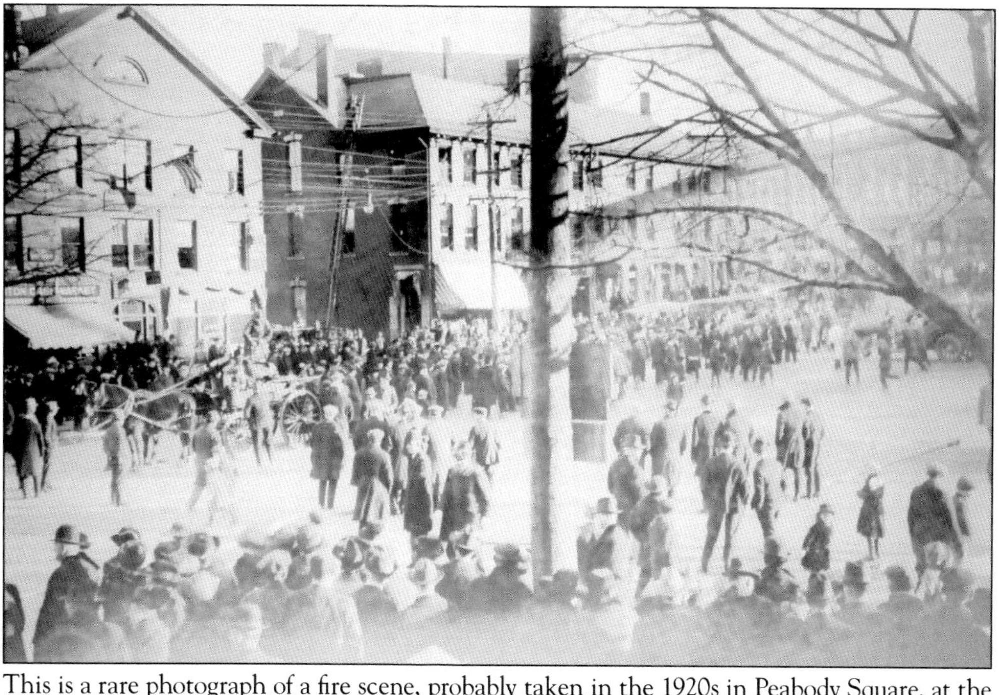

This is a rare photograph of a fire scene, probably taken in the 1920s in Peabody Square, at the Allen block. Notice the ladders extended on the left side of the building.

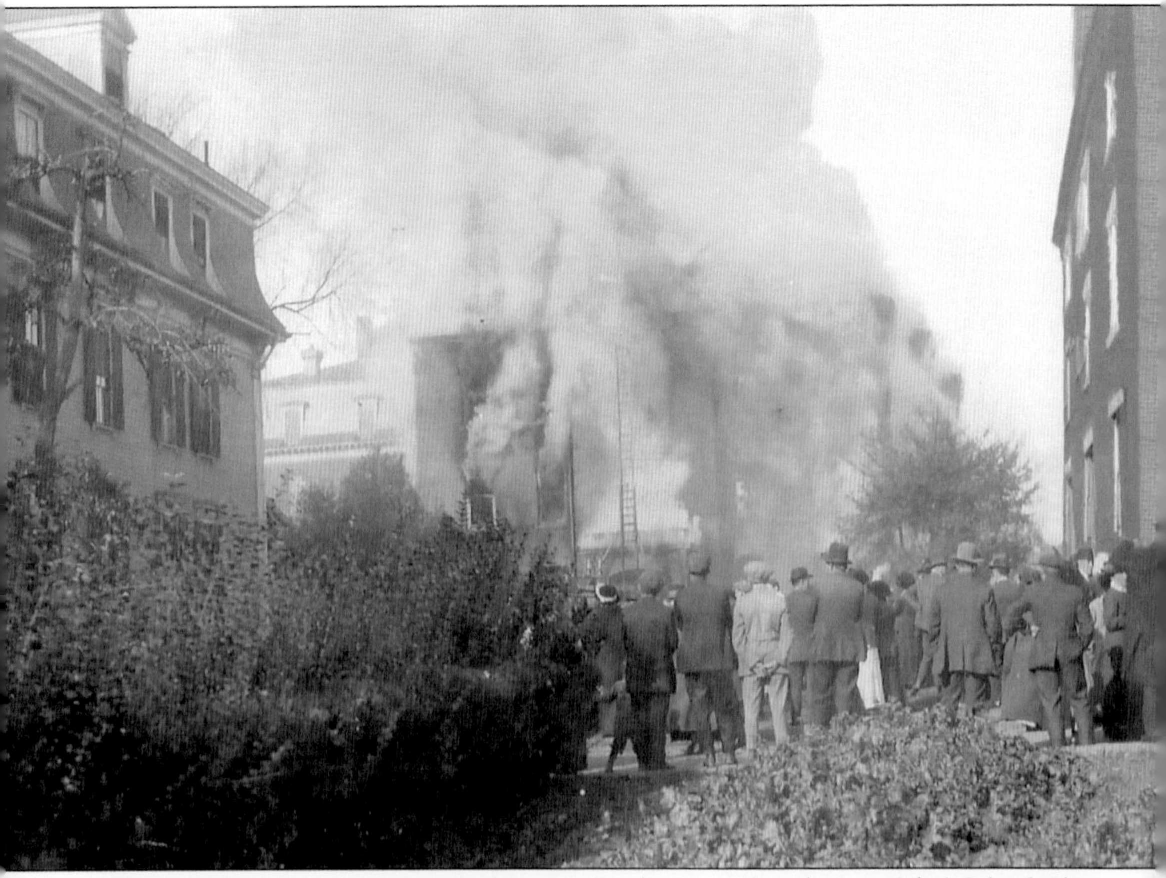

The most tragic fire to occur in Peabody was on October 28, 1915, at the Saint John's School. That day, 21 young girls lost their lives. Many photographers arrived from all over to get the best shots. The police at the scene confiscated many cameras and even locked up a few photographers for taking pictures when told not to, which makes this photograph very rare. Firemen were helpless in this fire, which started in the basement of the school, traveled up the stairway, and blocked the exit for the children to leave safely. They panicked and still tried to exit. Most of the girls who were killed were right in this doorway, only feet from freedom. There were many tales of heroes and rescuers that day who helped to stop the rising death toll. It is truly amazing to learn that many people in Peabody, especially younger firefighters, are not aware of this tragedy.

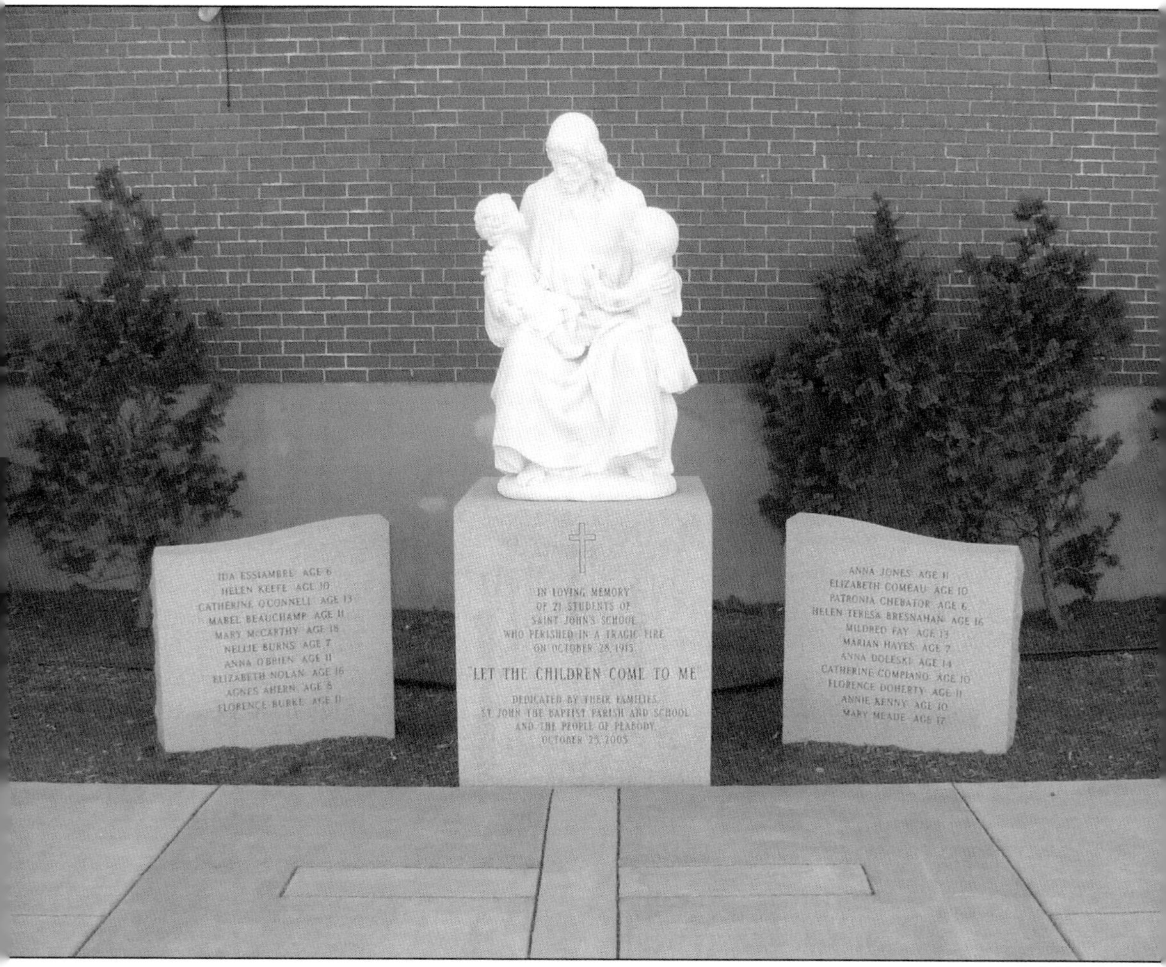

In October 2005, almost 90 years to the day, a memorial was dedicated to the girls who were killed in the fire at the Saint John's School: Helen Keefe, 10; Agnes Ahern, 8; Elizabeth Nolan, 16; Anna O'Brien, 11; Elizabeth Comeau, 10; Anna Jones, 11; Helen Teresa Bresnahan, 16; Mary McCarthy, 18; Mildred Fay, 13; Florence Burke, 11; Florence Doherty, 11; Catherine Compiano, 10; Mabel Beauchamp, 11; Nellie Burns, 7; Anna Doleski, 14; Marian Hayes, 7; Mary Meade, 17; Ida Essiambre, 6; Annie Kenny, 10; Catherine O'Connell, 13; and Patronia Chebator, 6.

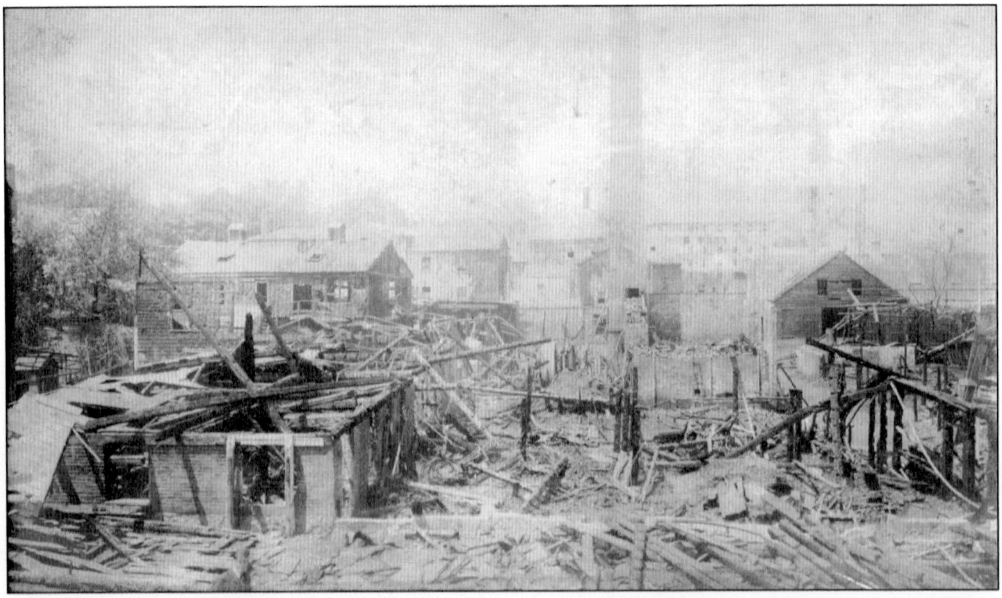

A familiar sight in Peabody was the ruins of yet another tanning yard or leather factory. Once a fire took hold of a building, soaked with chemicals and with its open space construction, the building was all but doomed.

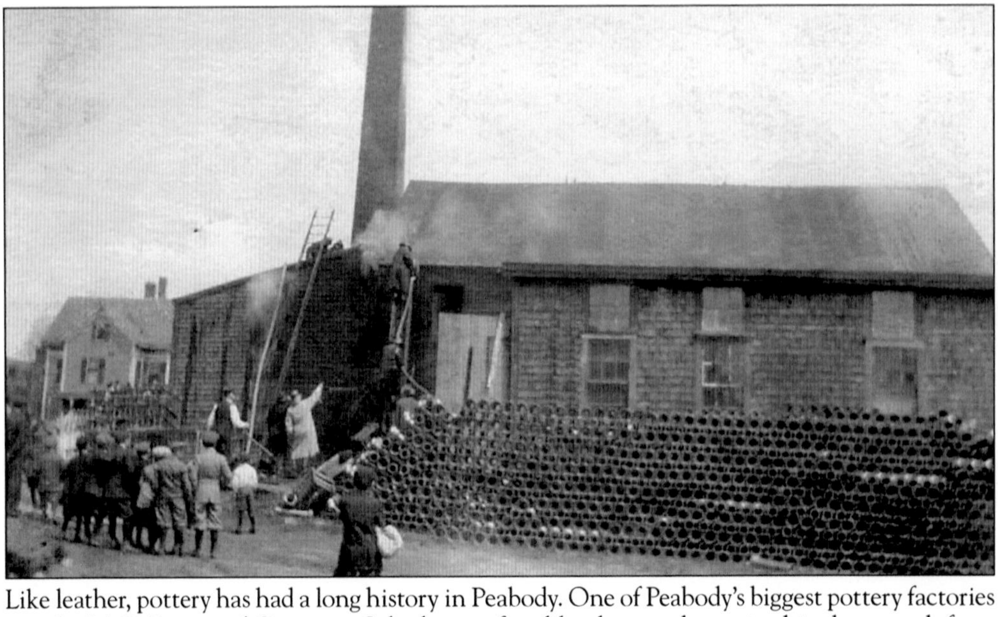

Like leather, pottery has had a long history in Peabody. One of Peabody's biggest pottery factories was the M. B. Paige and Company. It had many fires like the one shown in this photograph from about 1915.

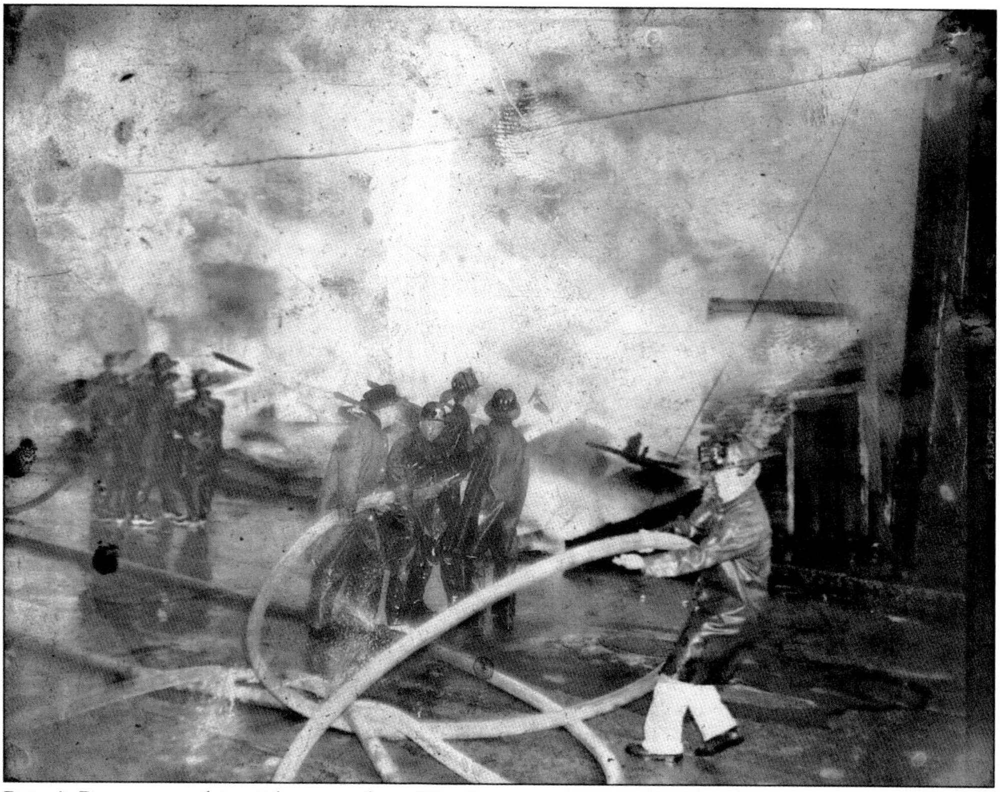

Paige's Pottery was hit with many fires. The last two, one on V-J Day (Victory over Japan Day) in August 1945 that almost destroyed all the buildings and nearly caused the destruction of the Wilson Square area and the second one in 1954, put an end to the long history of pottery in Peabody.

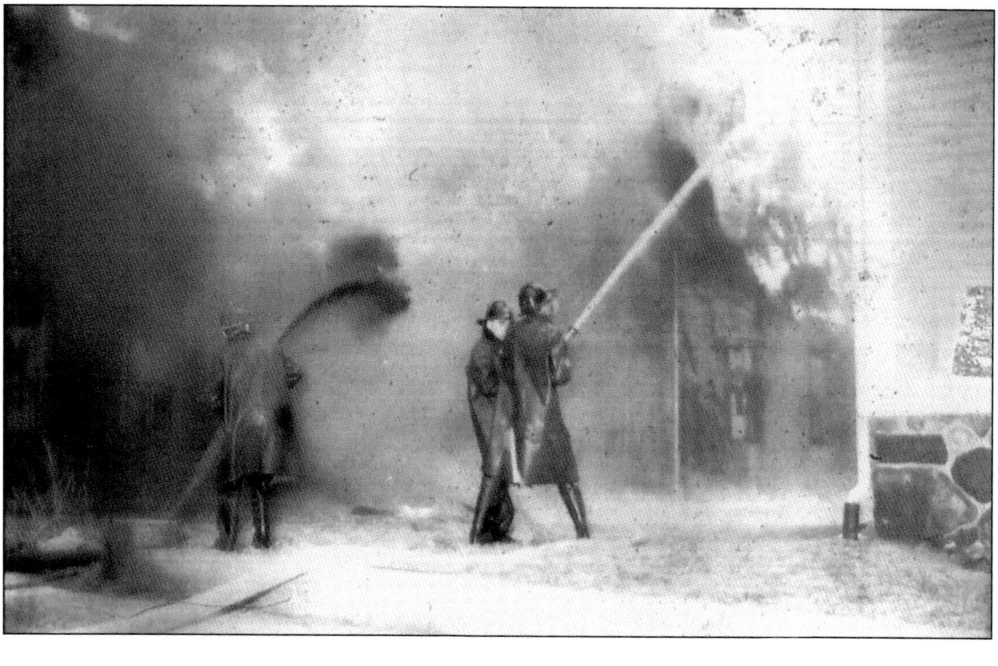

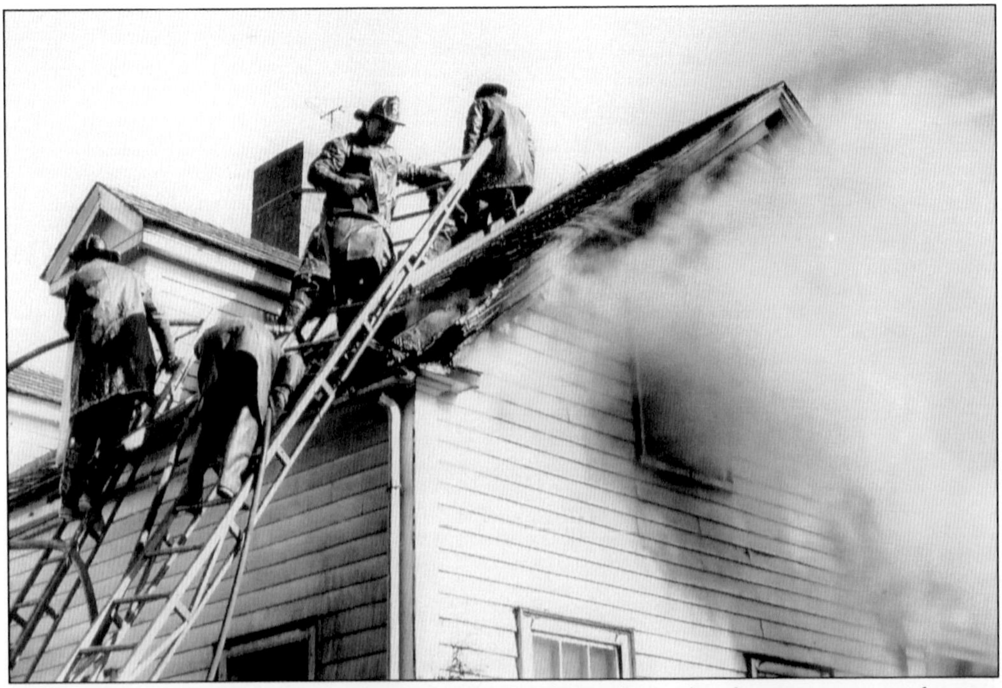

Shown here are action shots at two unknown fires where the firemen were performing their duties.

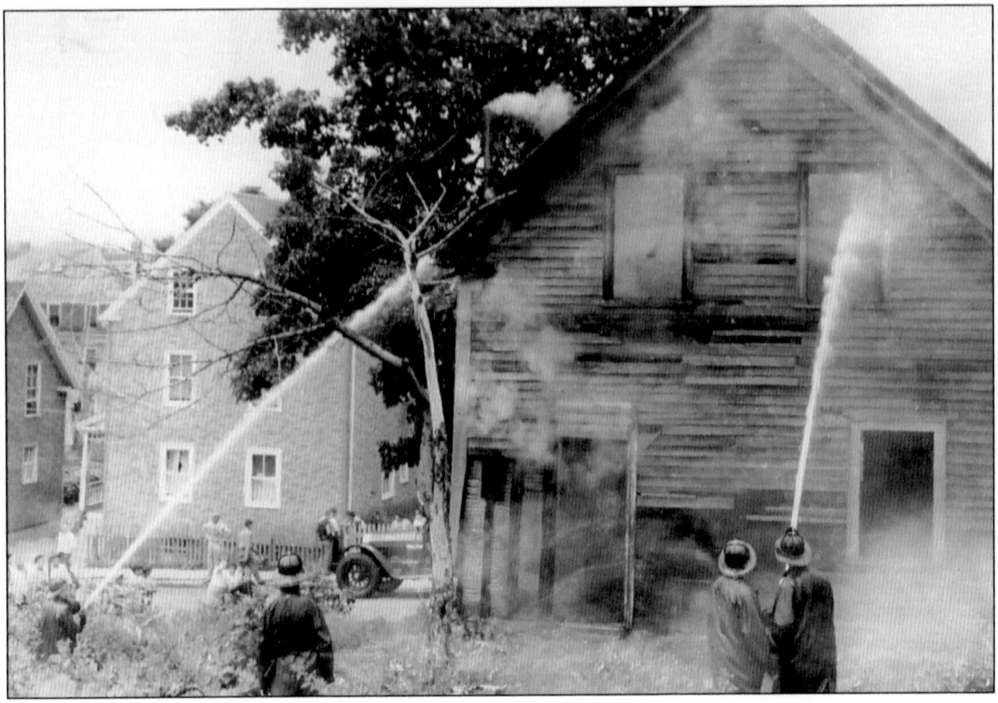

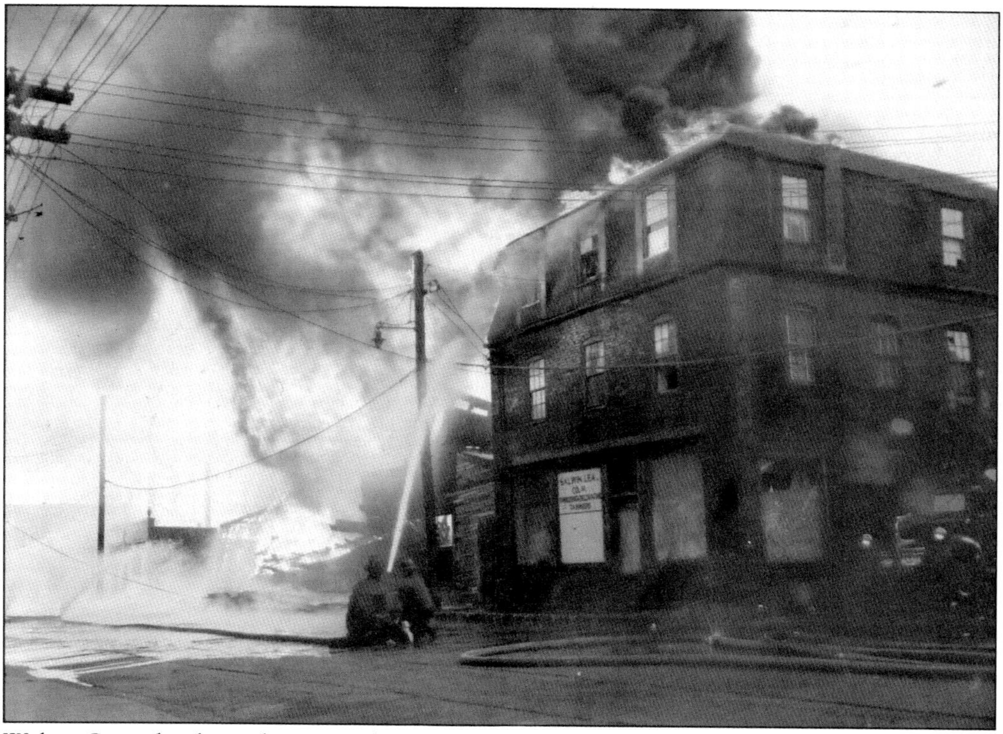

Walnut Street has been the scene of many multiple-alarm fires in Peabody, including this one on Saturday, August 4, 1951, at the M. W. Ellis Grain and Salwin Leather Companies, which caused $300,000 to $500,000 worth in damage and put many people out of work.

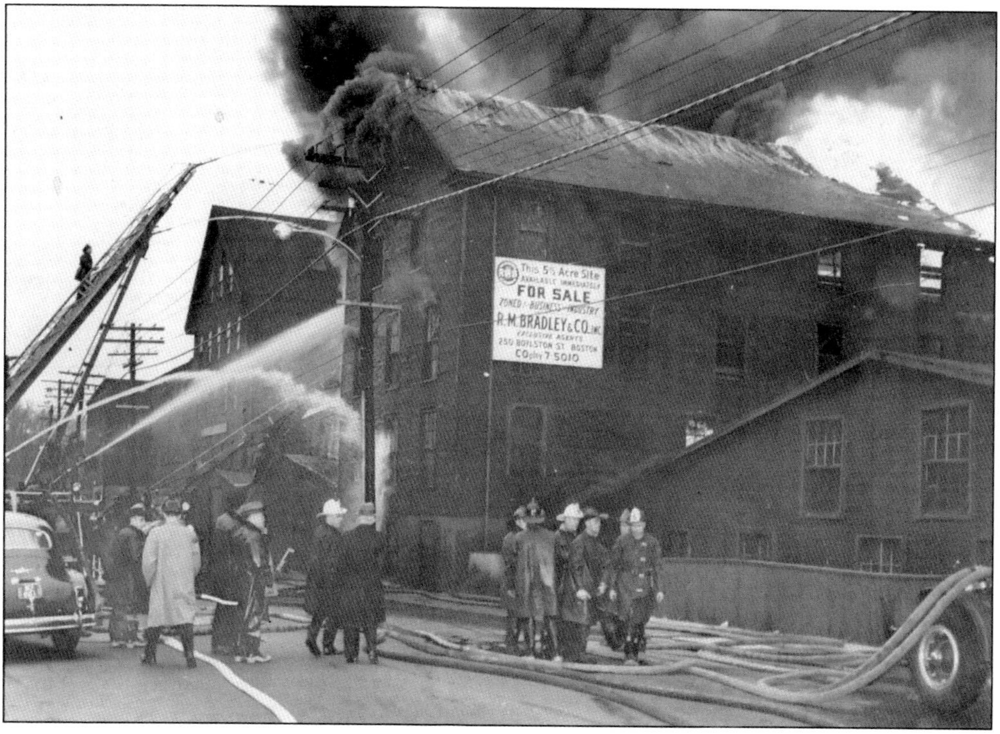

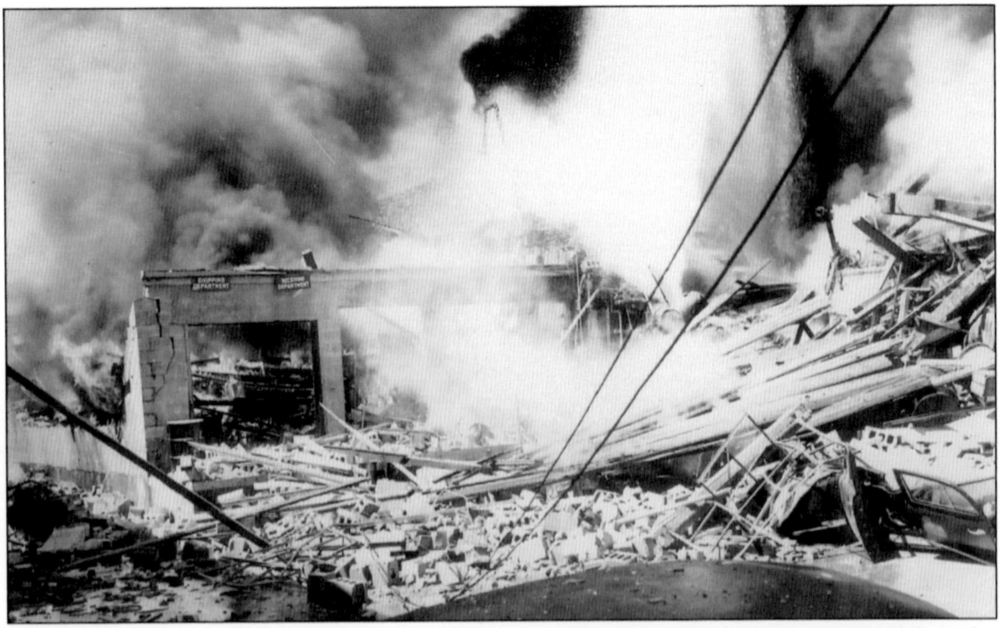

Peabody has had its share of explosions caused by chemicals, whether they were used in the tanning industry or for glues and adhesives. One of the worst explosions was on April 30, 1953, at the American Polymer Corporation on Foster Street, which resulted in the loss of just one life. The devastation at the scene can be compared to the November 2006 explosion at a chemical plant in Danvers.

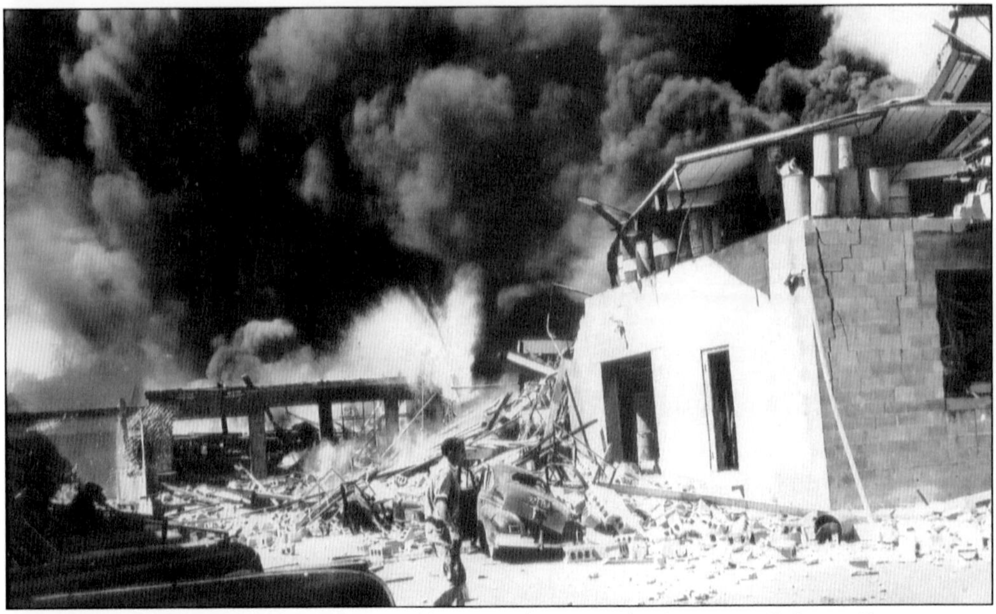

Notice the carnage in this picture caused by an explosion at the American Polymer Corporation: cement blocks, destroyed automobiles, leaking chemicals, and live electrical wires. The burning chemicals belched deadly black smoke and gases and caused extremely dangerous air quality. The barrels that were still intact in proximity of the heat caused concern because of the potential for more explosions called BLEVEs, or boiling liquid expanding vapor explosions.

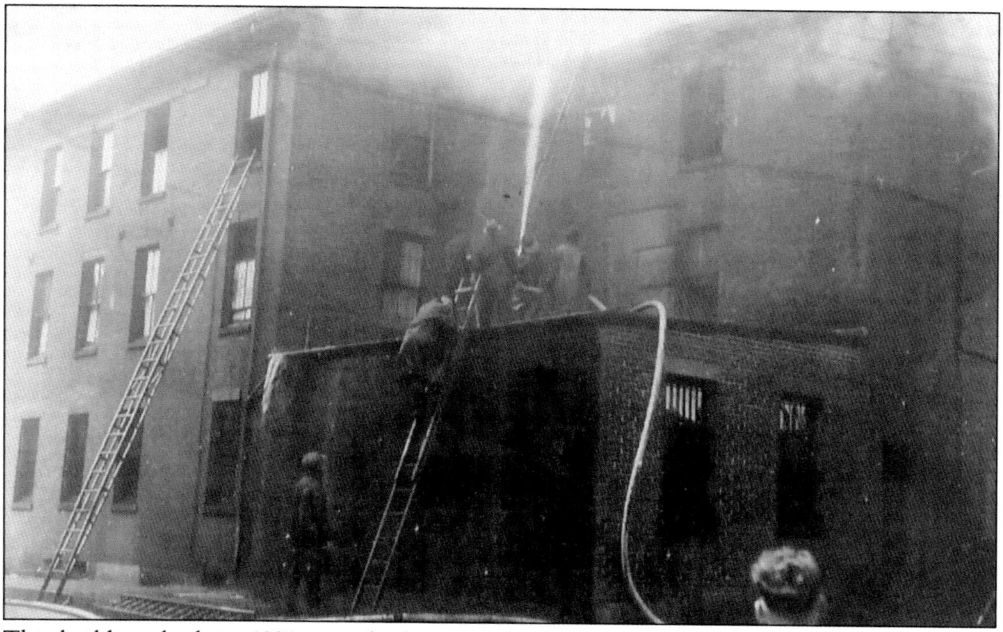

This building, built in 1828, was the home of the Dustin Hotel located at the corner of Mill and Central Streets. It escaped the great conflagration of 1843 but has had other fires, including the one depicted here and another in the 1960s that demolished the third floor, leaving it as a two-story building. The Century Bank now occupies the building, and it is also the location of the old Heifer pub.

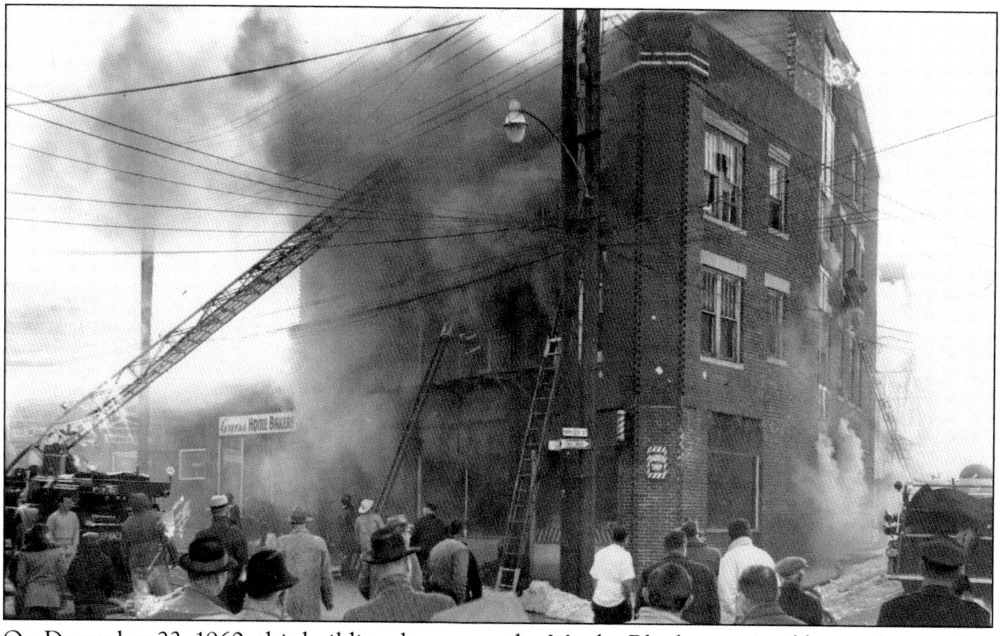

On December 23, 1960, this building, known as the Mooka Block, sustained heavy damage due to a fire. It was remodeled, but on February 27, 1989, there was a second Mooka Block fire that destroyed the buildings to the left. Motor supervisor Ernest Nelson, who was acting captain at the scene, was removed from the second-floor hallway and, along with firefighters Nicholas Gerakaris and Ray Meserve, was sent to the Josiah B. Thomas Hospital for injuries.

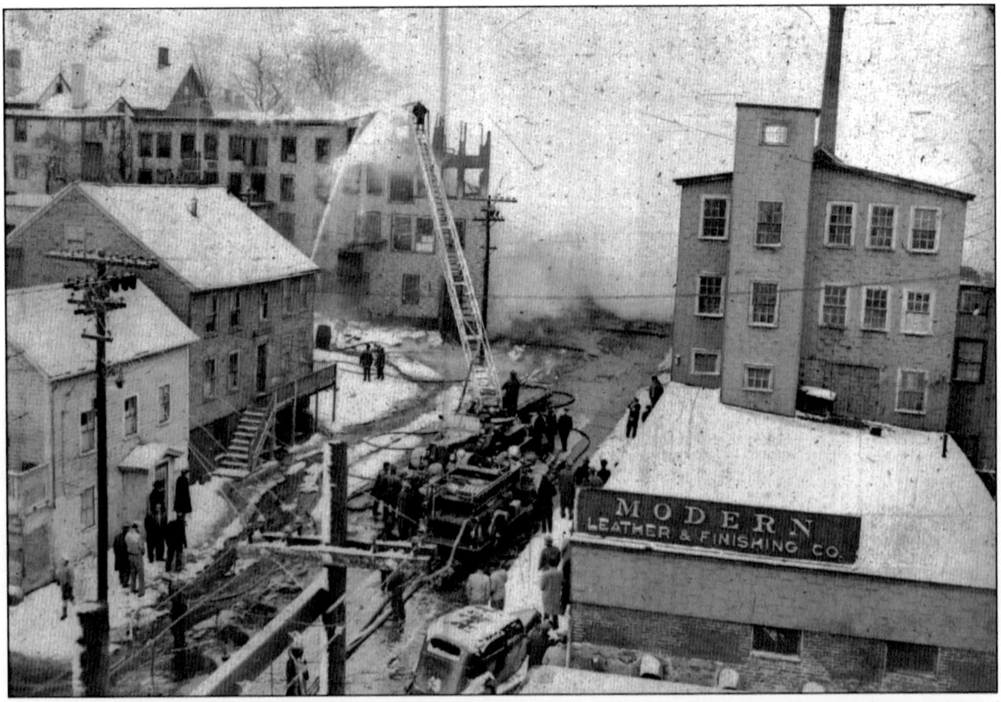

This fire at Modern Leather and Finishing Company, located on Spring Street at the corner of Park Street, was on January 7, 1952. It destroyed most of the building.

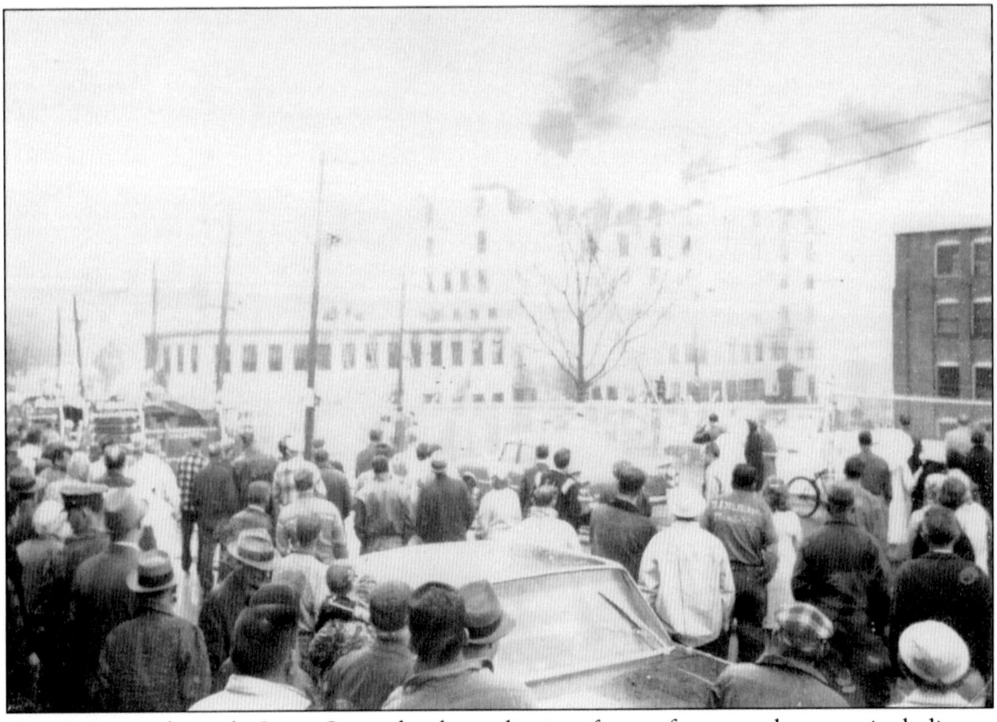

Howely Street, formerly Grove Street, has been the site of many fires over the years, including at the Gnecco and Grilk Company, Flynn Leather, and Southwick.

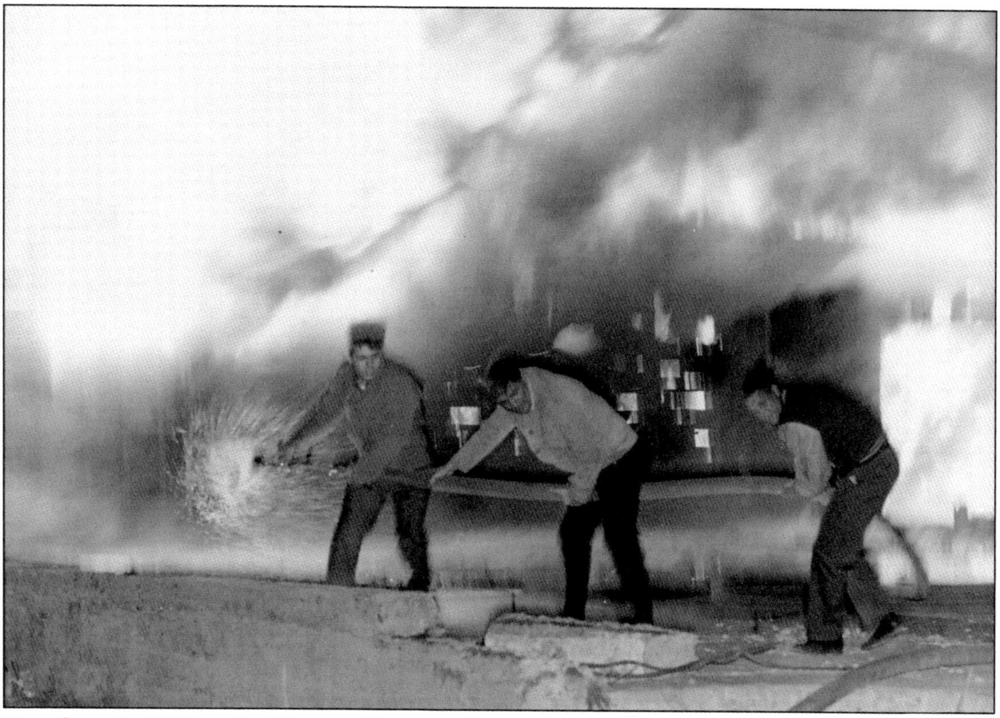

Seen here in 1967 at the Kirstein Tanning Company fire, firemen are ducking while trying to avoid the heat from the flames, which might have been creating its own wind. Notice the spray on the hand line.

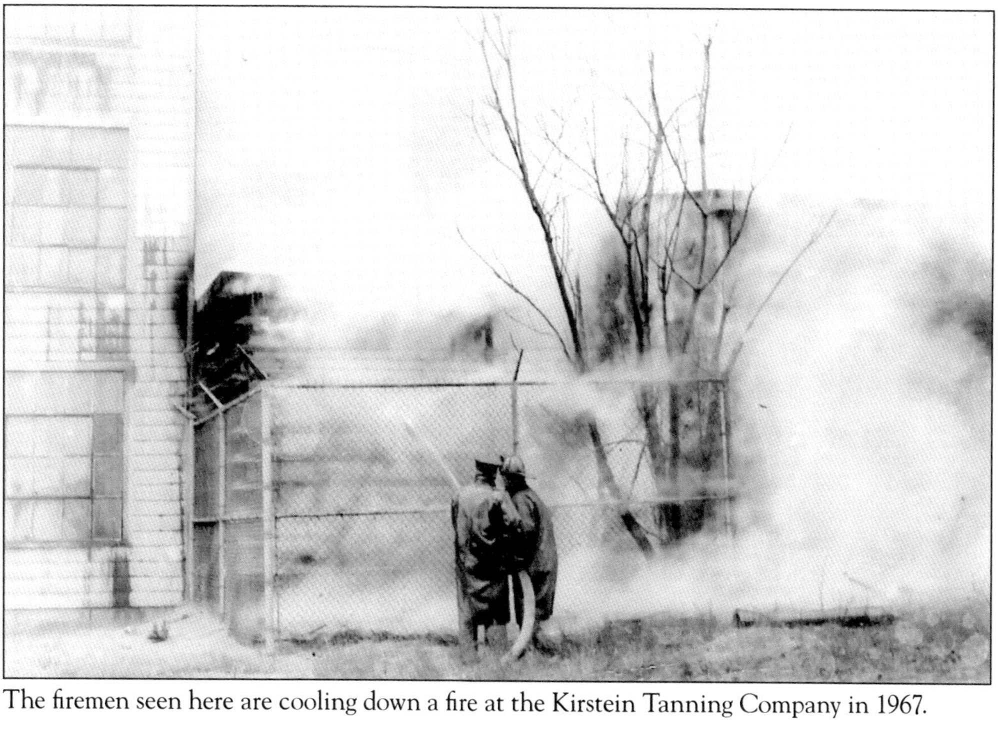

The firemen seen here are cooling down a fire at the Kirstein Tanning Company in 1967.

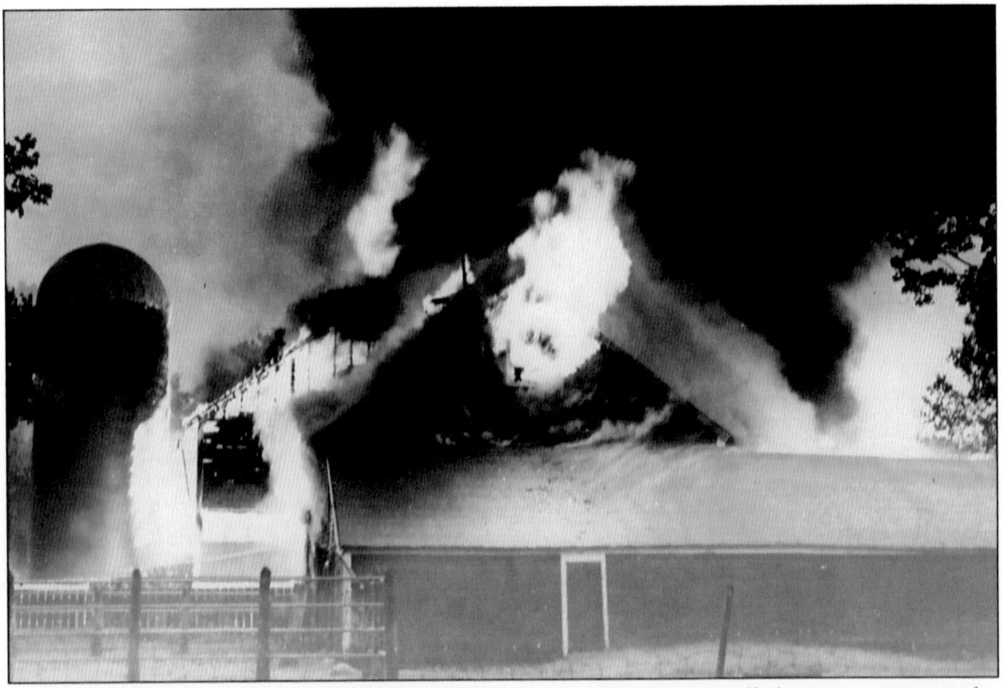

A barn fire at an unknown location burns out of control. Protecting all the exposures was the strategy of this fire.

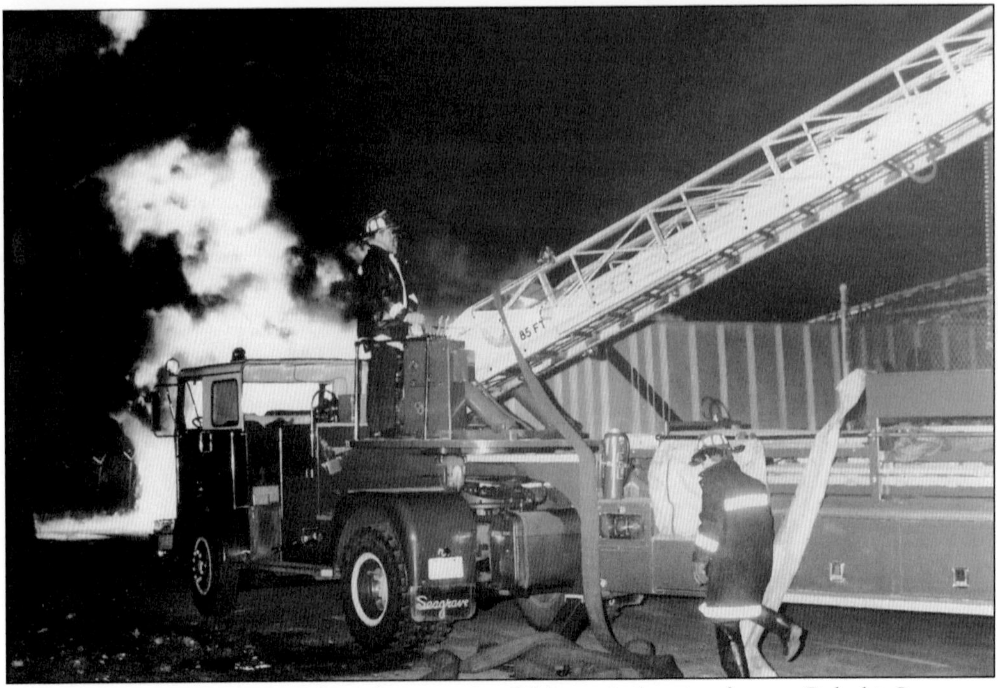

The ladder crew is seen working here at the Webster Industries fire on Pulaski Street on January 28, 1975.

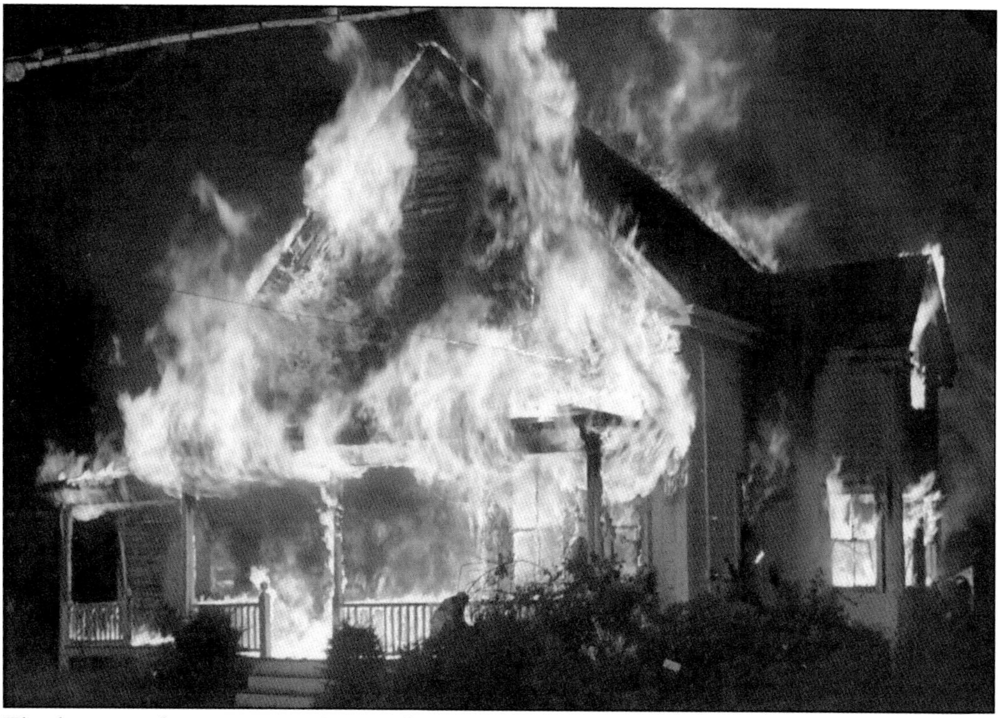

This house explosion was at 464 Lowell Street on July 16, 1973, and was fully involved upon the first engine company's arrival.

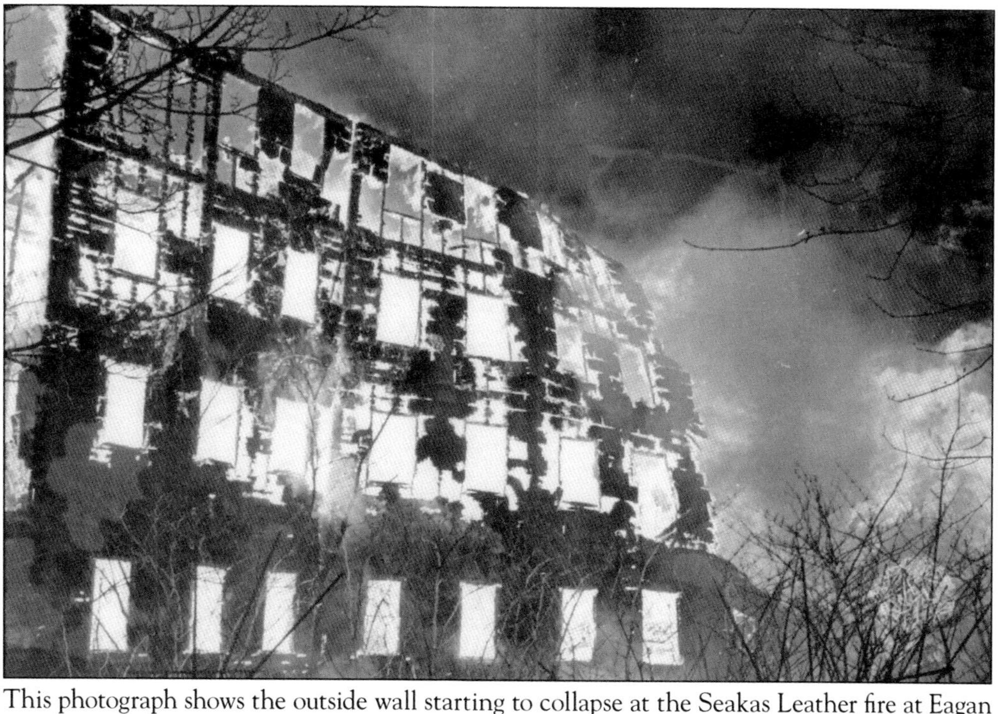

This photograph shows the outside wall starting to collapse at the Seakas Leather fire at Eagan Place, also known as Barrel Alley.

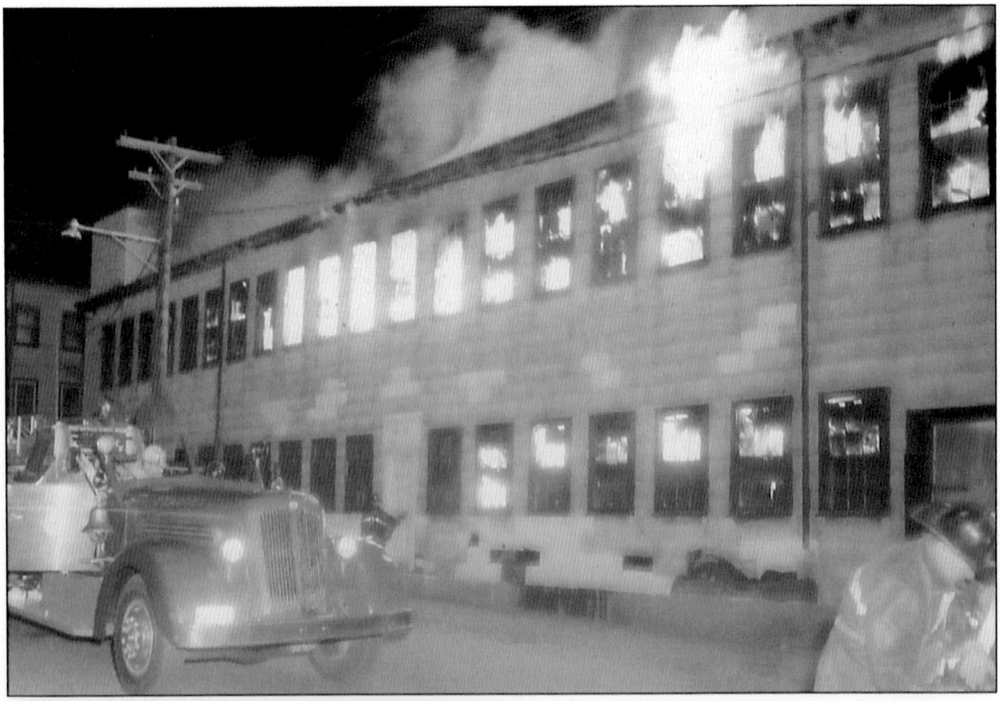

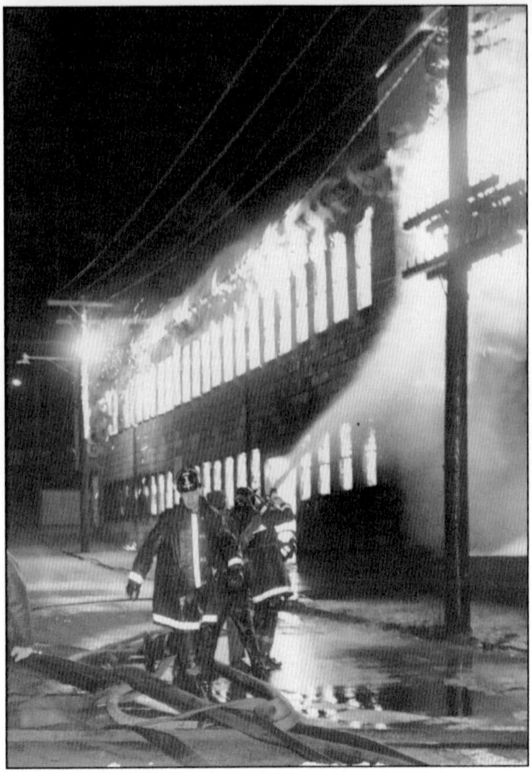

These photographs show yet another factory fire in Peabody, this time at the Gnecco and Grilk Company on Howley Street. Firefighter John Kerwin is seen in the picture on the left maneuvering around the many hoses laying about.

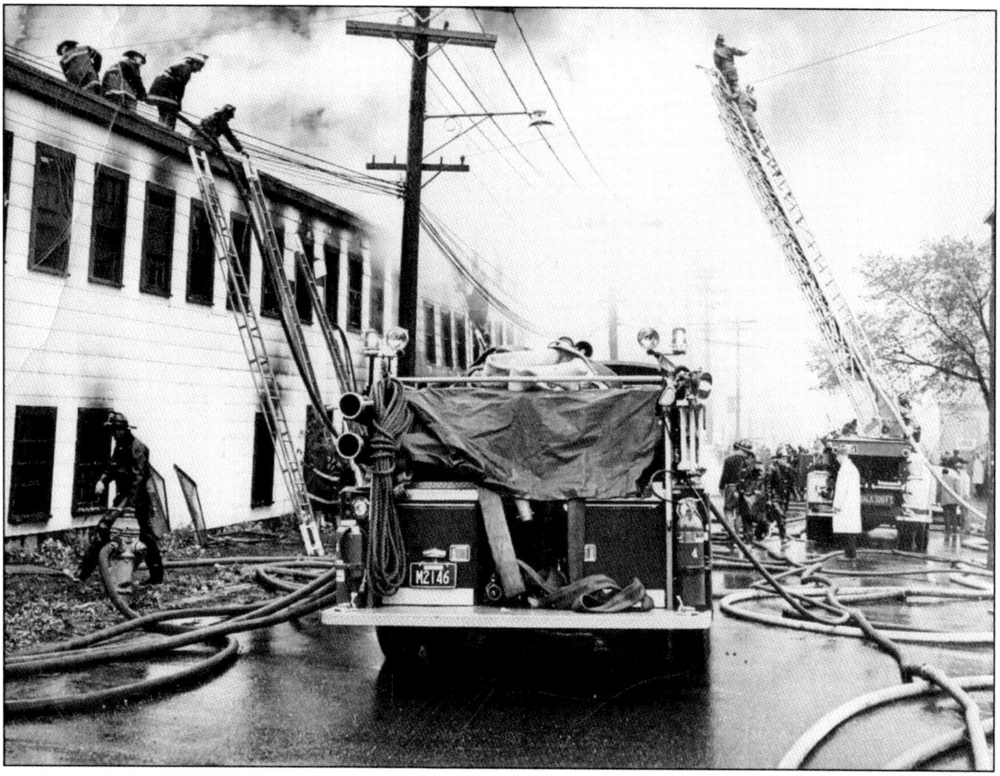

This emergency room photograph is of an unknown worker from the Gnecco and Grilk Company fire on May 14, 1964. Gnecco and Grilk Company was a leather shop located on Howley Street. This fire took the lives of five workers. The firefighters found some of the men in lockers and under desks.

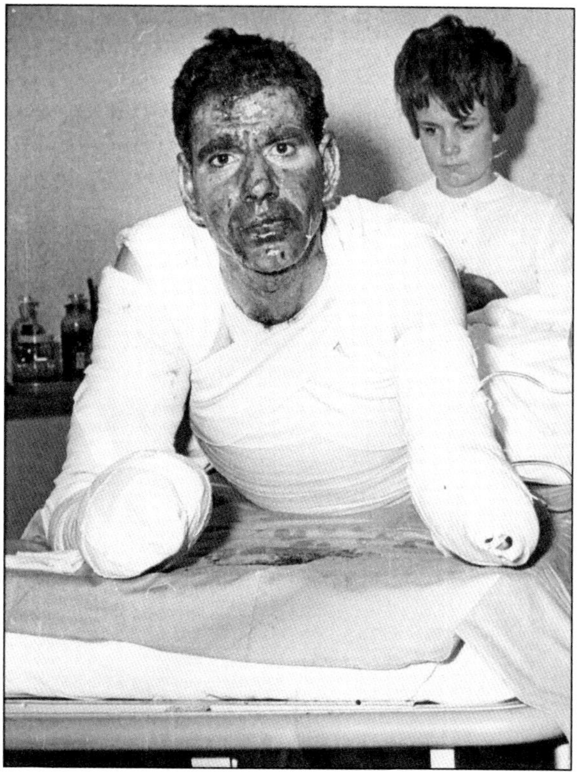

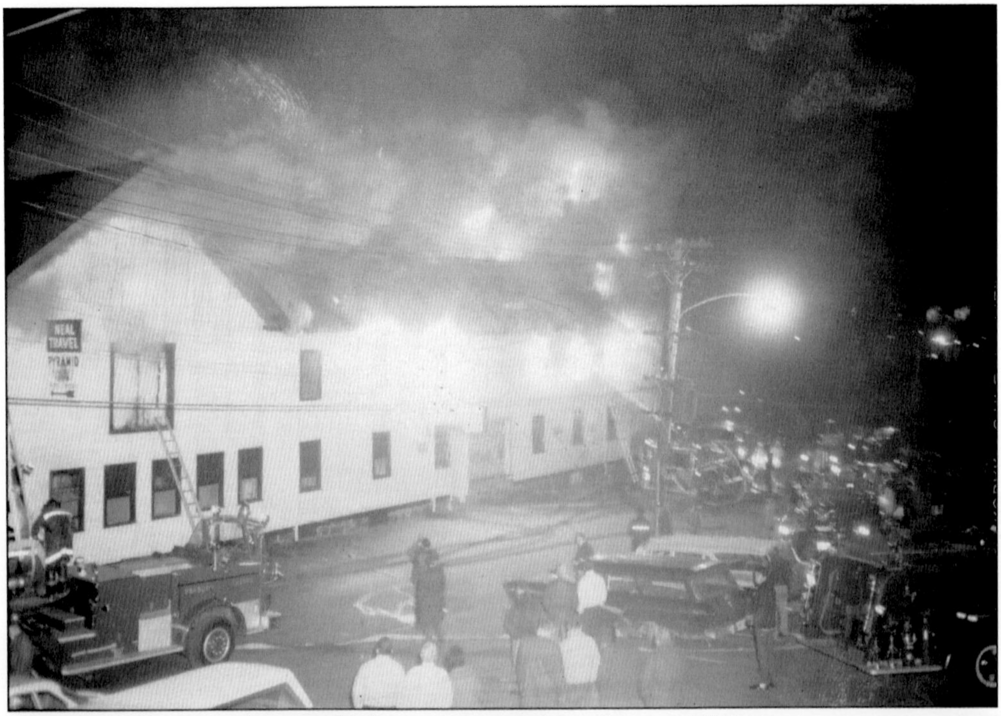

This December 1967 photograph shows a general-alarm fire on Mill Street at a warehouse and Neal Travel. This is the current location of the Charles Motel.

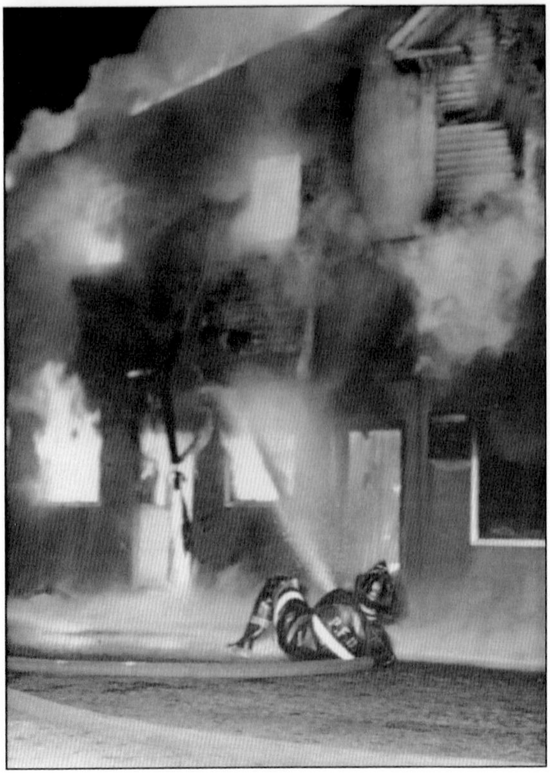

This photograph shows Fred Trueira trying to control a hand line at a fire on Walnut Street.

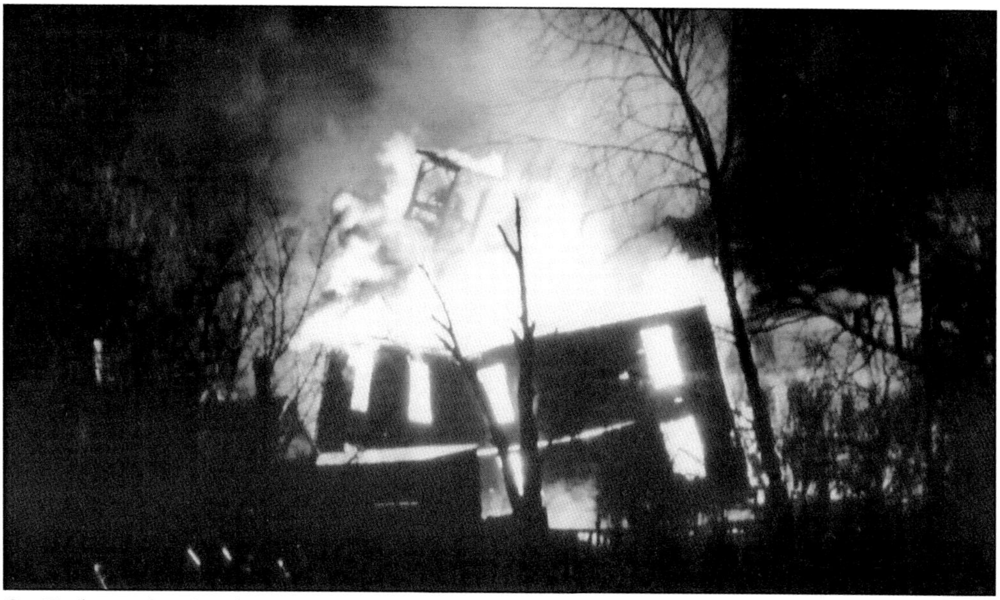

On Friday night, April 14, 1961, at 2:05, box 18 was pulled at the engine No. 5 firehouse. It was pulled by one of the firefighters, and it was no false alarm. A fire had started somewhere in the station, and the two firefighters who were stationed there had to leap out a second-floor window to escape the fire. The next day, a body of a transient person who had snuck in from the cold was found in the ruins. The cause was never known. The only thing that was salvaged from the ruins was the fire bell that hung in the bell tower.

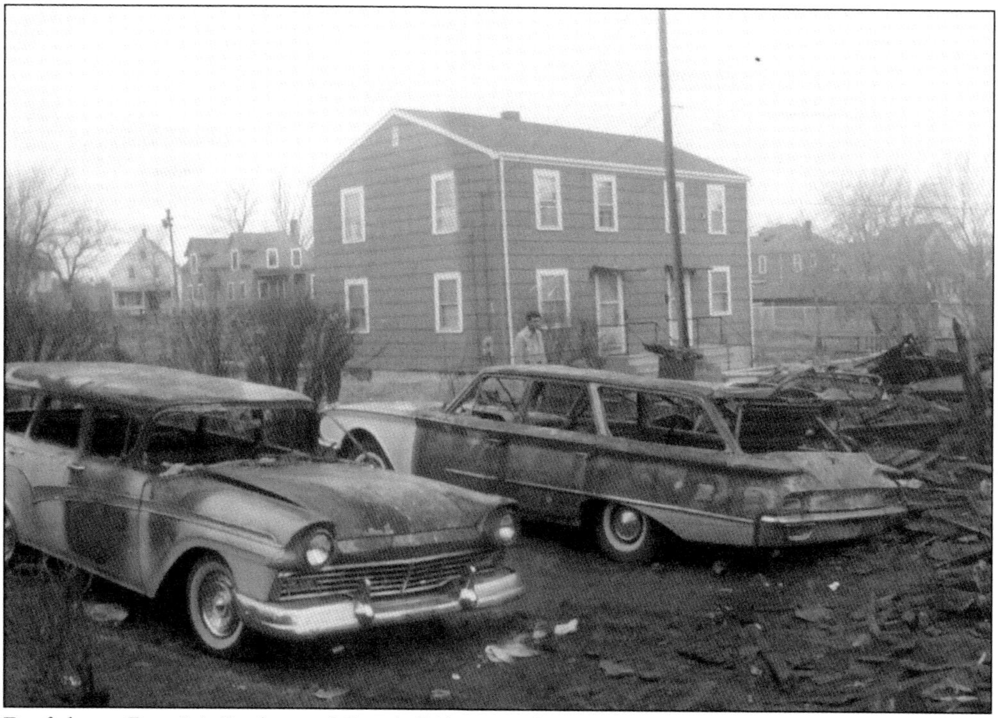

Firefighters Don McCauley and Frank D'Amico, the two firefighters who escaped the engine No. 5 firehouse fire, had to walk home from work that day due to their cars being destroyed.

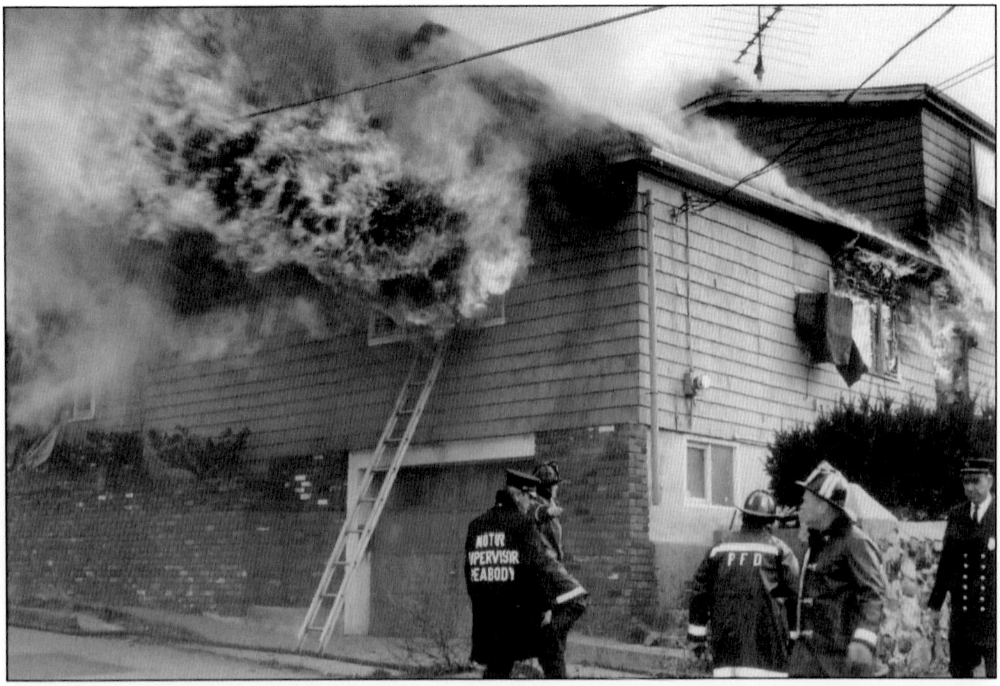

This photograph shows fire blowing out the window at 39 Fulton Street on December 11, 1974. Deputy Peter Pappas (in dress uniform), motor supervisor Arthur Blake, Bucko McCauley (second from right), and the rest of the men figure out the best strategy for putting this fire out.

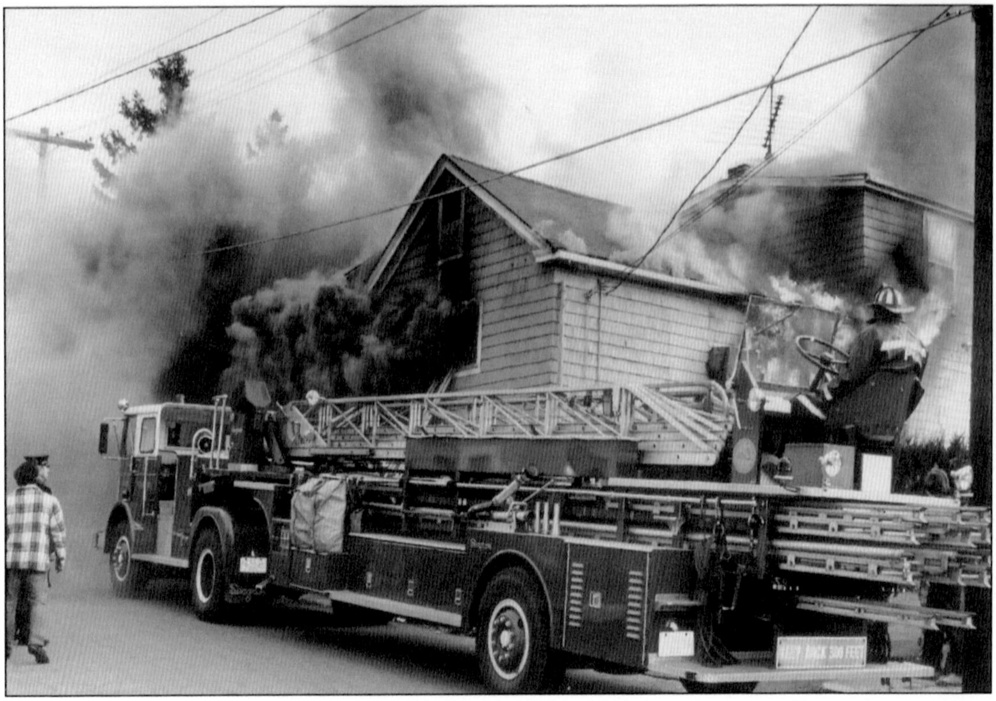

In this photograph, ladder No. 1 is pulling up to a fire at 39 Fulton Street on December 11, 1974. On the tiller, firefighter Peter Fallon awaits orders.

Firefighters William "Bones" Mahoney (No. 3), Steve Markarian (foreground), and others render aid to an exhausted man that they have saved, as Capt. Henry Velez catches his breath at a fire at the Tannery Apartments at 12 Crowninshield Street. The Tannery Apartments were once part of the A. C. Lawrence Leather Company and were renovated in the 1980s into apartments.

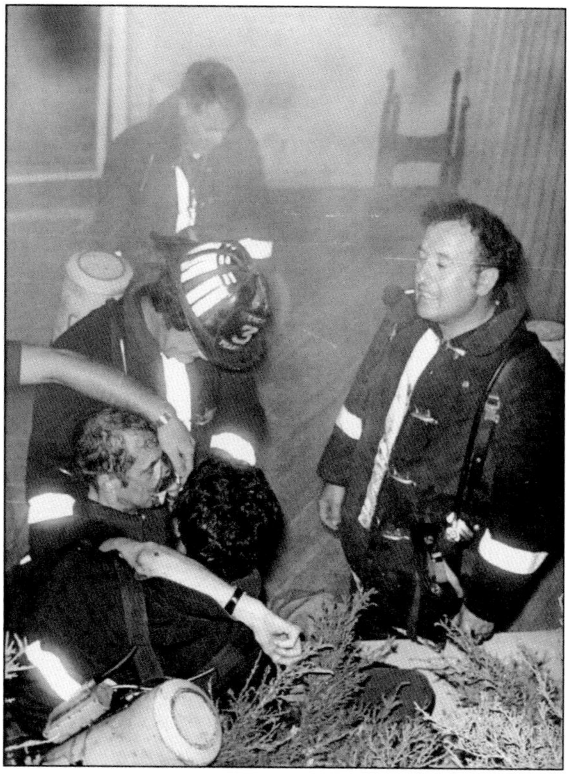

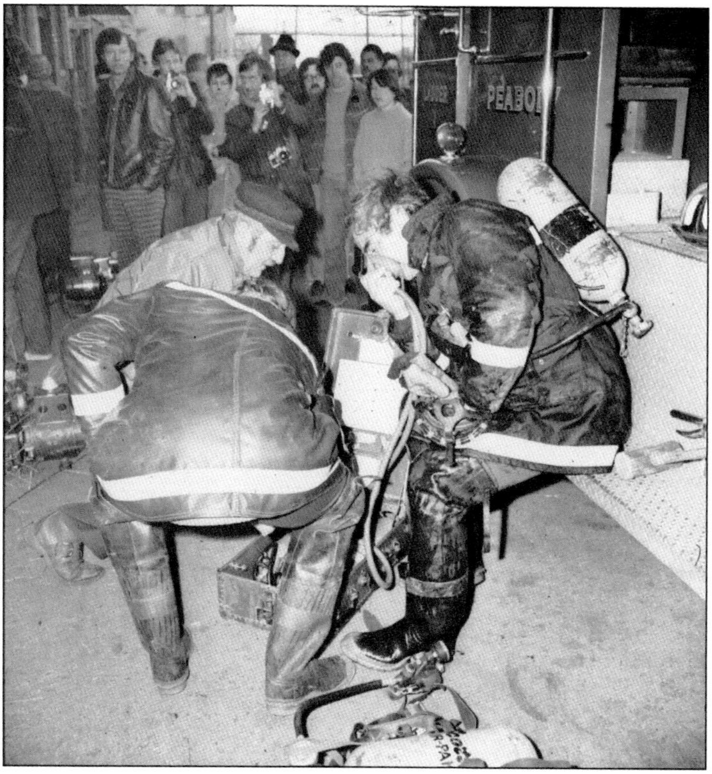

Deputy Chief Tony Koutzoukis (left) and an unknown firefighter administer oxygen to firefighter Dan Strabone (right) at a fire at the Q building of the old A. C. Lawrence leather factory on 12 Crowninshield Street on February 2, 1975.

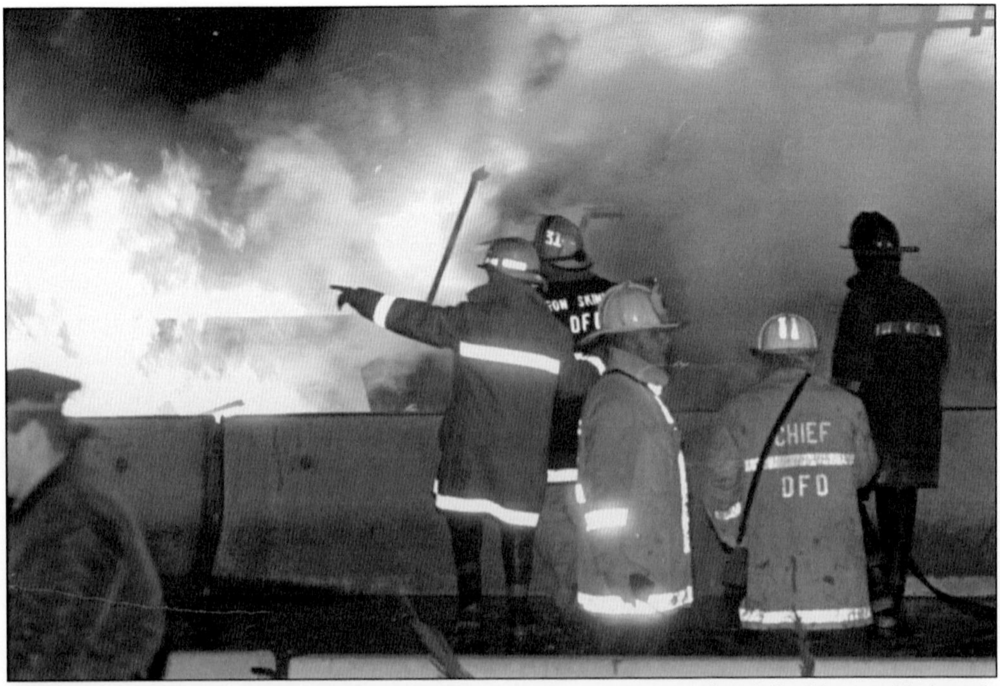

This fire was on Route 128 northbound on December 2, 1985. A tanker full of flammable liquids struck the Jersey barriers and ripped open its tank causing a massive explosion and fire that had to be extinguished with firefighting foam.

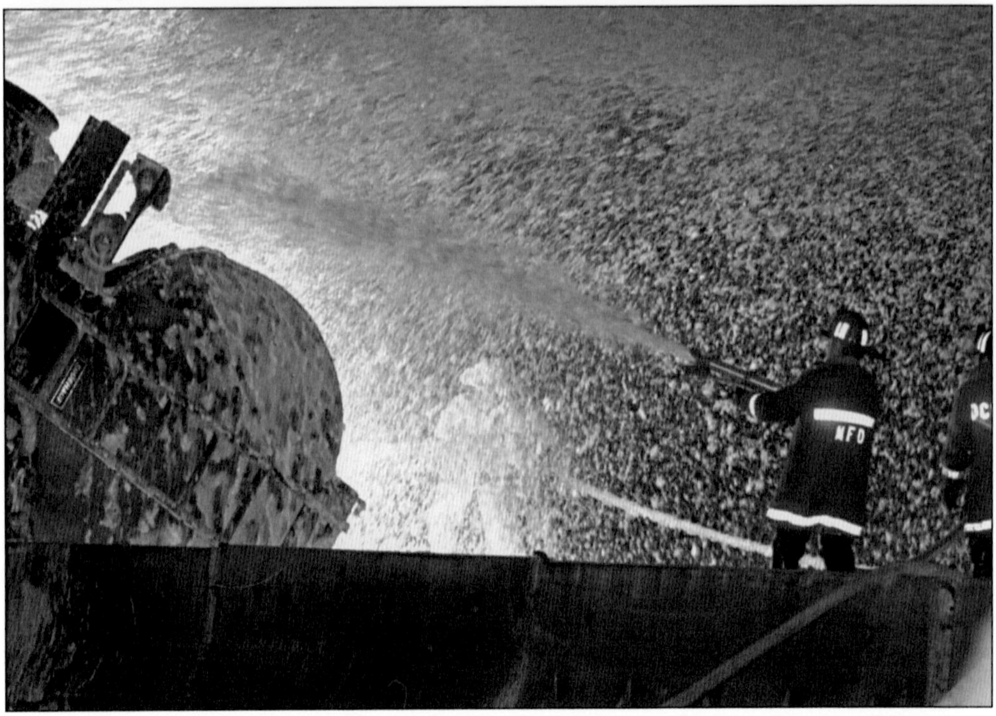

Looking like snow, the foam used to fight this fire helped smother the flames and kept it in check. Route 128 was closed for several weeks because it had to be dug up and repaved.

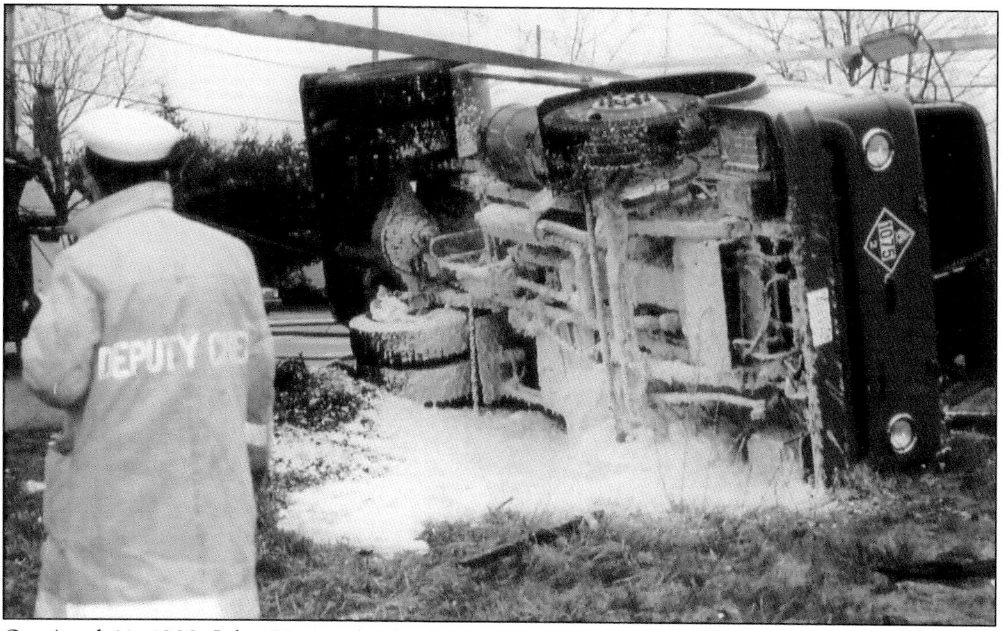

On April 11, 1988, John H. Smith, the driver of this propane tanker, died from chest trauma when the propane tanker rolled over at the corner of Route 114 and Mount Pleasant Street. The tanker was owned by Suburban Propane out of New Hampshire. The call came in from the police at 2:09 p.m. A one-mile area was evacuated. Foam from Beverly's engine No. 8 with Chief Ken Pelonzi, Lynnfield's engine No. 1, and Middleton's engine No. 1 was used to prevent an explosion.

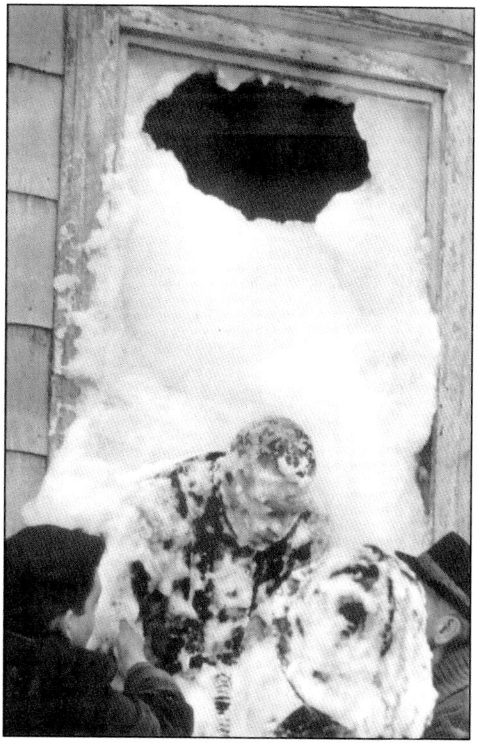

Shown here is Donald "Bucko" McCauley being pulled from a window of a house on Paleologous Street that was being used for a firefighting foam demonstration that went bad. The demonstration almost took his life and the lives of Jack Mullarkey, Arthur Blake, and Ray Brown.

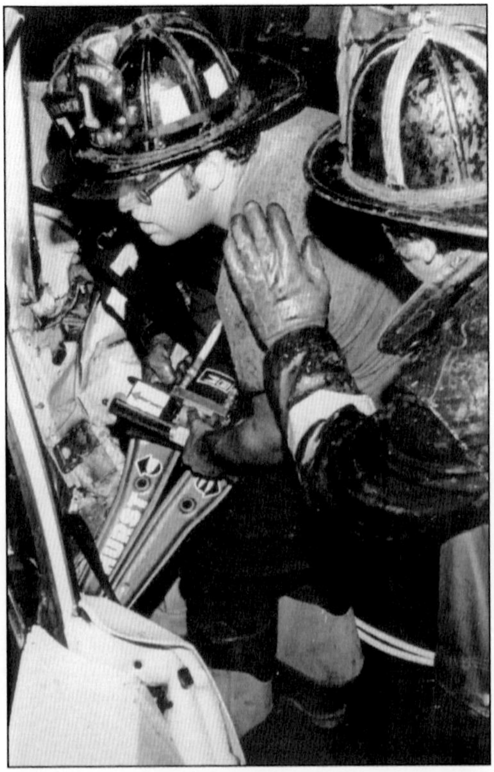

Peabody's first set of hydraulic lifesaving tools that firefighters used to cut victims out of mangled cars, known as the Jaws of Life, was donated by the Peabody Elks. Although the Jaws were not set to be unveiled and put into service on July 24, 1973, they were used for the first time on a car accident, seen in the picture on the left. Four youths survived the crash and proved the effectiveness of the new tool.

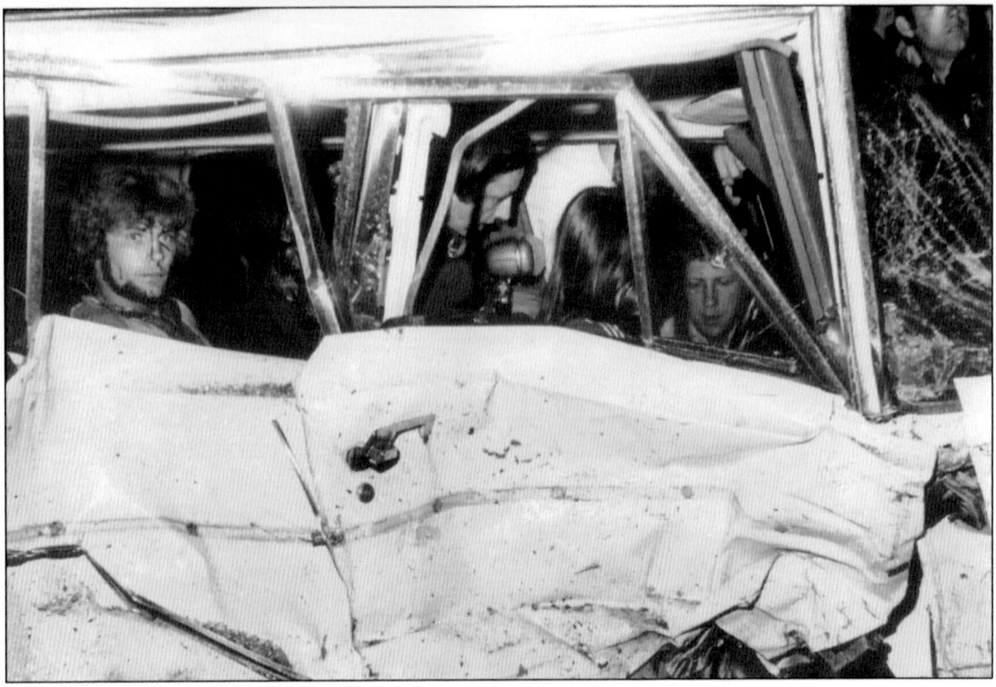

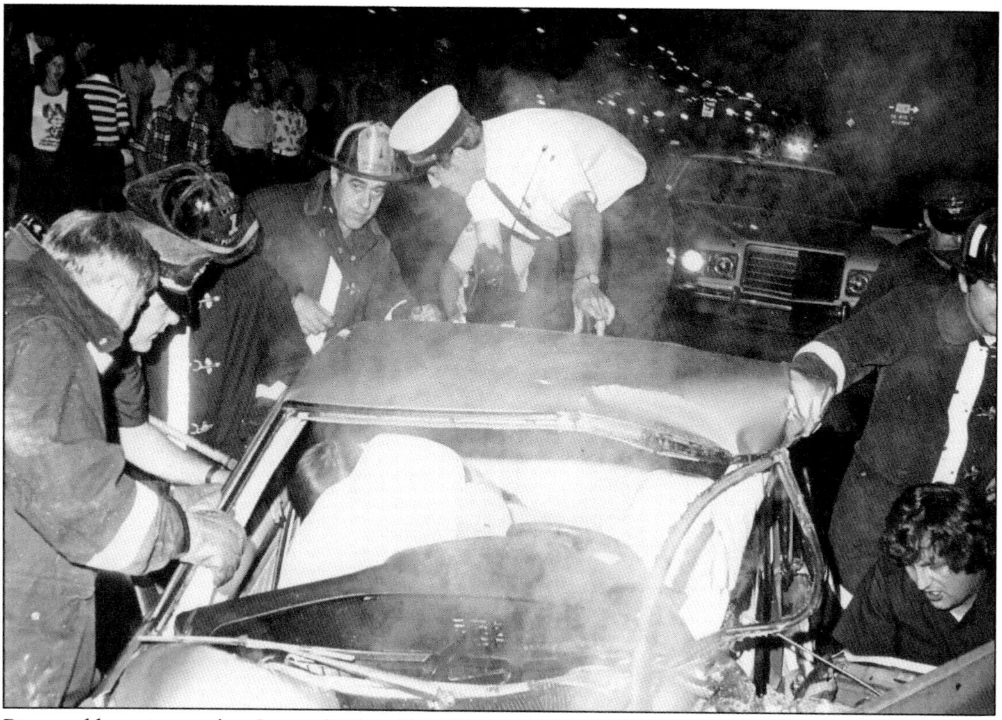

Pictured here is another Jaws of Life call on Route 114 near the on-ramps of Route 128 in front of the North Shore Shopping Center, a scene of many accidents and fatalities. From left to right are Capt. Manny Mendonca (facing the camera), Deputy James McDonell, firefighter Mike Zuppio, and Kevin Lynch. Below, firefighter William Burke demonstrates the new Jaws of Life that were just purchased and donated by the Peabody Elks.

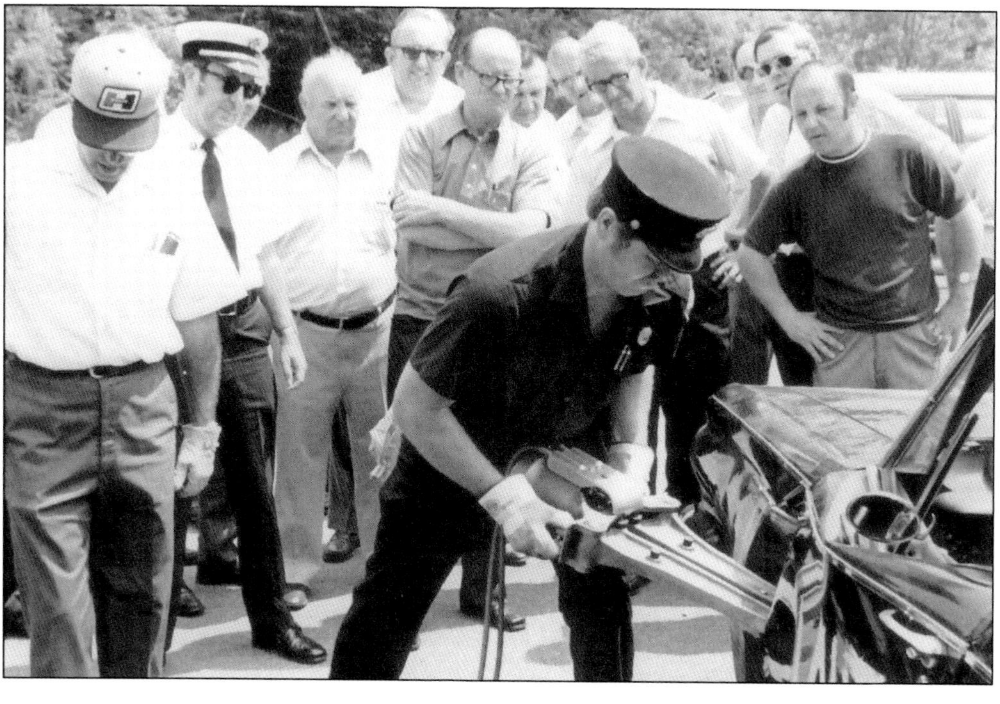

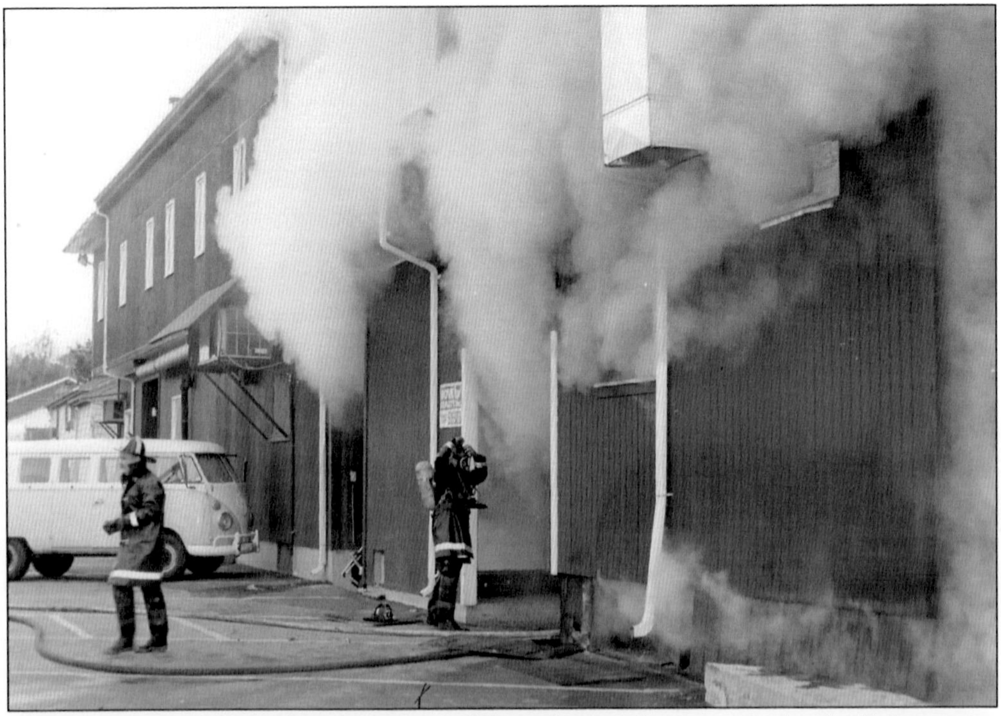

The Metro professional building at 484 Lowell Street burned three times before finally being torn down and rebuilt with masonry. This fire was on May 28, 1974, and was caused by an arsonist.

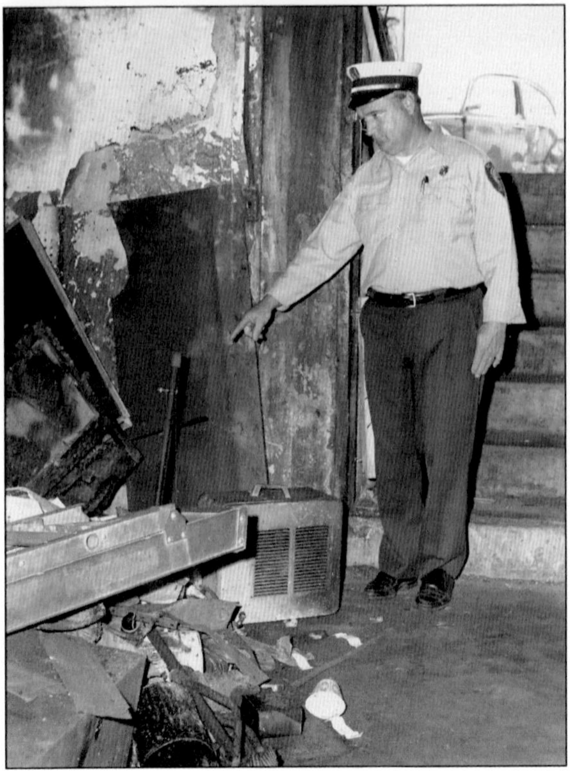

Fire prevention and investigating officer Deputy Frank Connors points to the location of were the charred body of the arsonist was found slumped against the basement door of the Metro professional building.

This fire on May 23, 1972, was at the South Essex Sewer District. Flammable liquids that had spilled into the waterway upstream from a fire the previous day caused the fire. A plumber's torch set off the ensuing explosion.

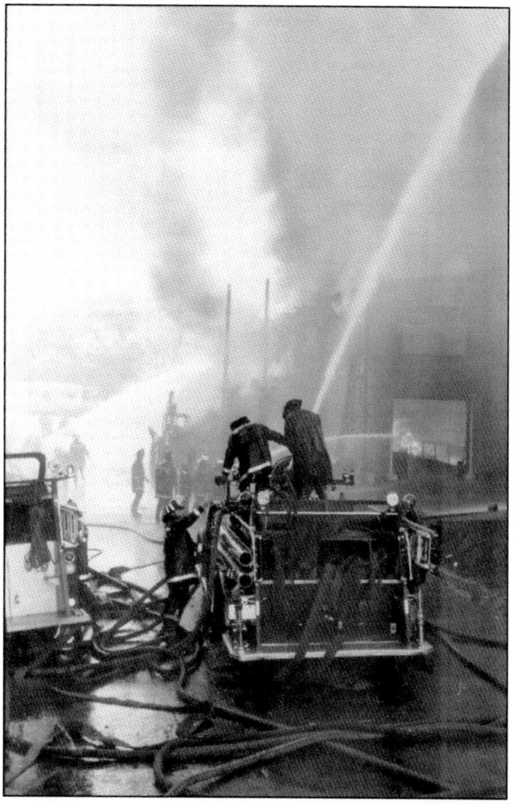

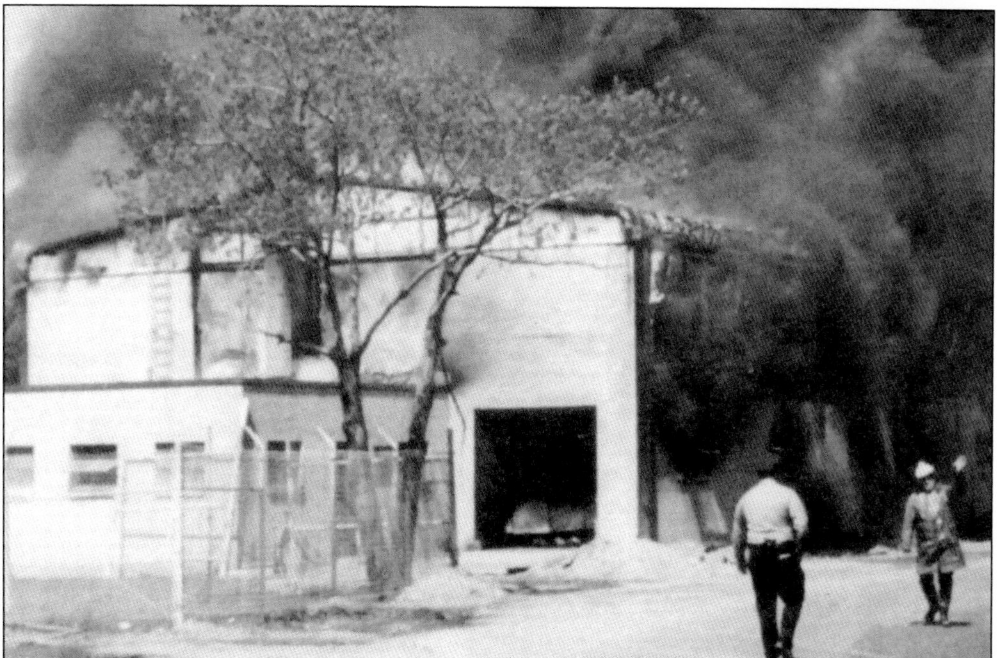

Deputy Lenny Lucey waves good-bye at his last fire, which was at the South Essex Sewer District. The fire used 10,000 gallons of foam to finally extinguish.

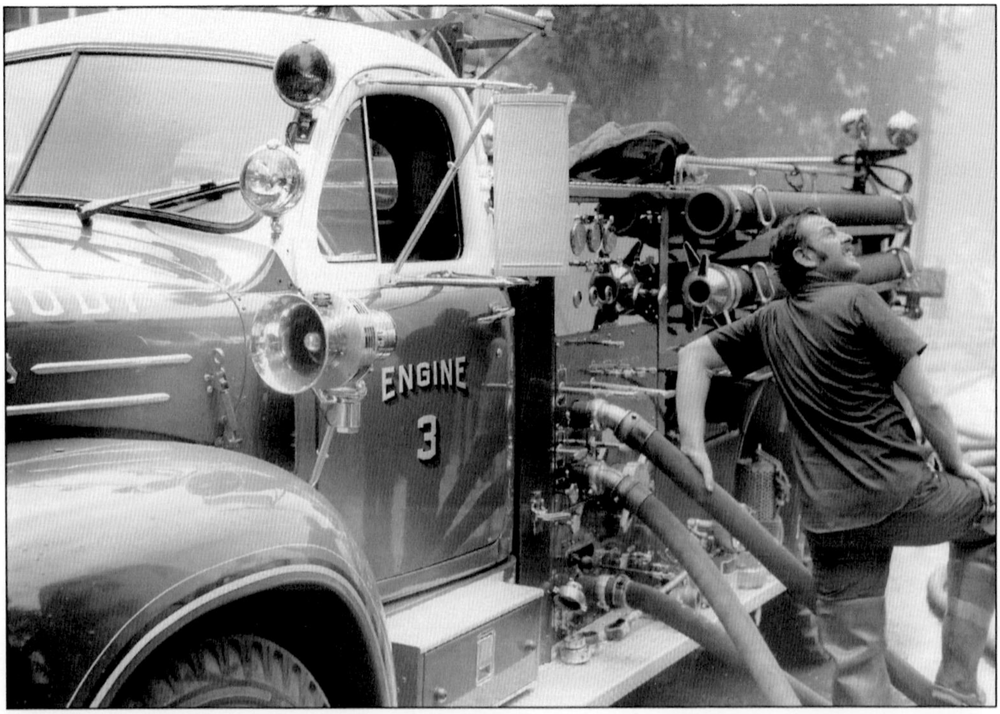

Firefighter Bill Downey runs the pump at the Warren school fire, on June 20, 1973. Downey retired in September 1998 as a captain in charge of engine No. 7 in West Peabody on Division 3.

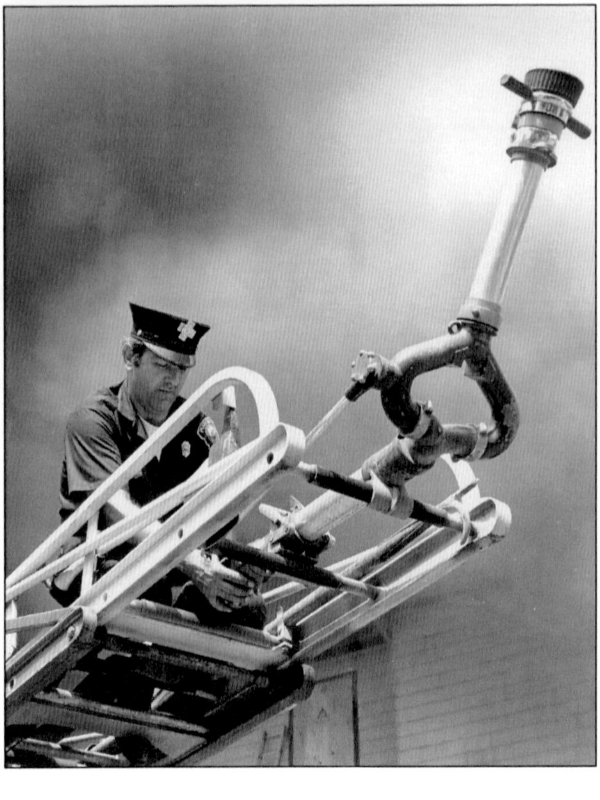

Firefighter Dan Strabone is seen here setting up the ladder pipe on ladder No. 1 at the Warren school fire on Buxton Street on June 20, 1973. Before they had the new aerial ladder trucks with the pre-piped waterway, firefighters used to have to hoist the nozzle onto the ladder bed and run the hose line up into the ladder bed.

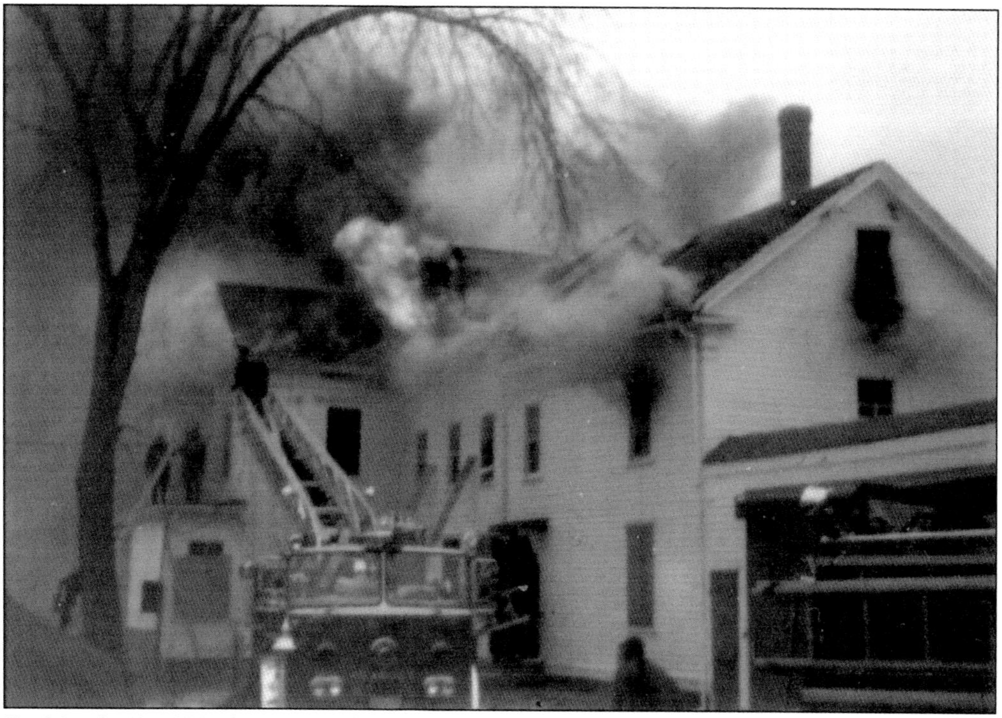

On March 23, 1973, there was a fire at the Crystal Springs nursing home on Goodale Street in West Peabody. The nursing home, which was abandoned at the time, was the old farmhouse of the Carten Farm. The Carten Farm was one of the biggest diary farms back in the day.

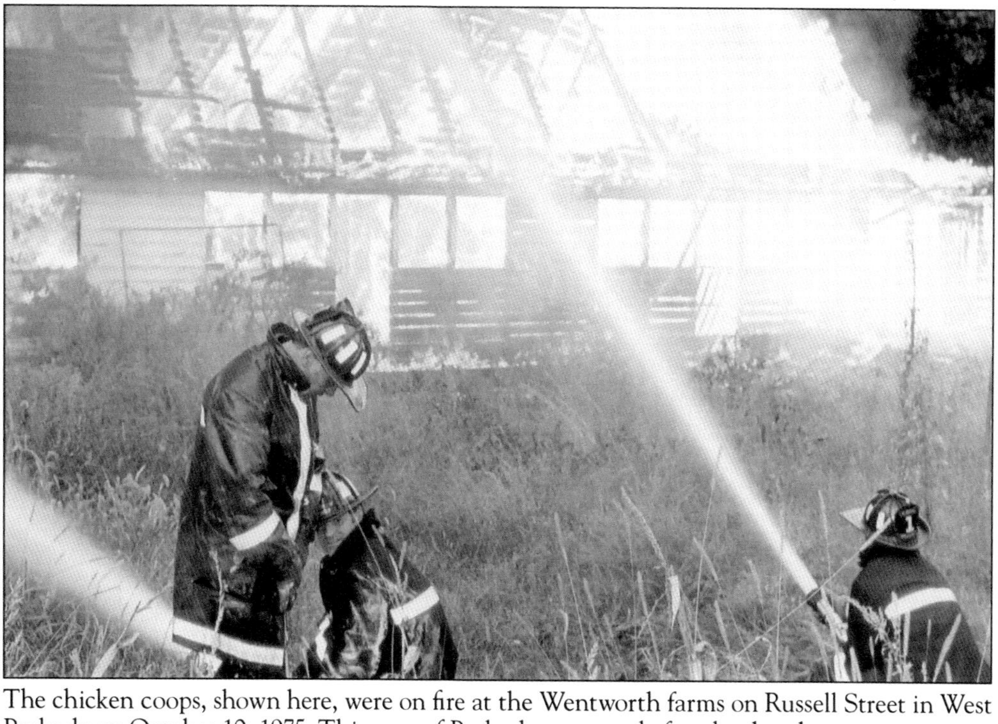

The chicken coops, shown here, were on fire at the Wentworth farms on Russell Street in West Peabody on October 10, 1975. This area of Peabody was mostly farmland and pastures.

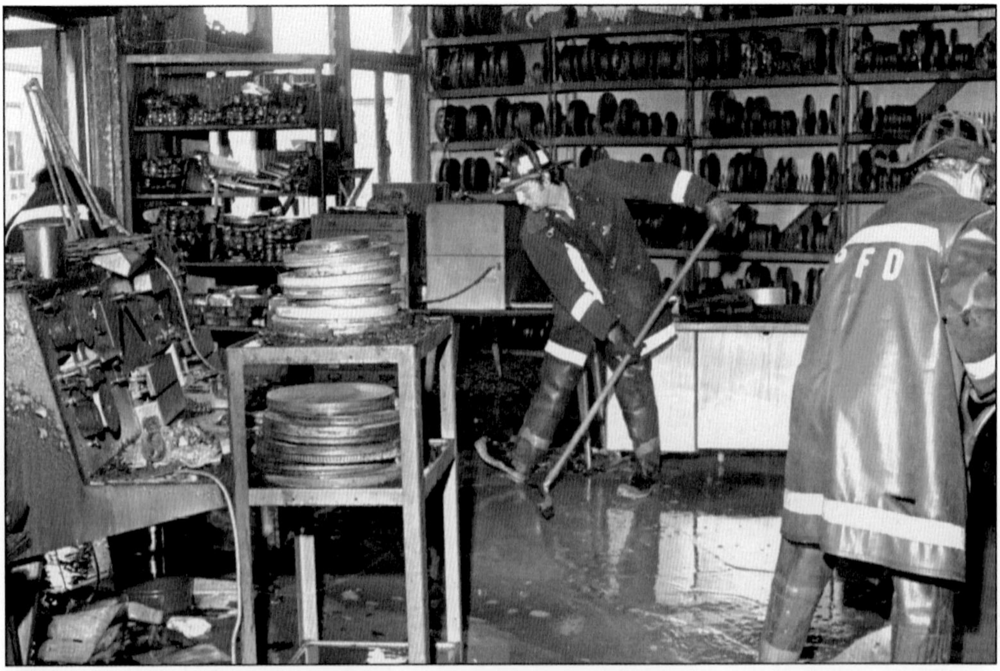

Within three years, the two middle schools in Peabody had caught fire, causing extensive damage. The first fire was at the Higgins Middle School (above) on King Street Extension on May 14, 1976. It started in the library and caused over $1 million in damage. The second fire was at the Kennedy Middle School (below) on March 23, 1979. It started in the general laboratory area and also caused a considerable amount of damage at a high cost. Both were blamed on arson.

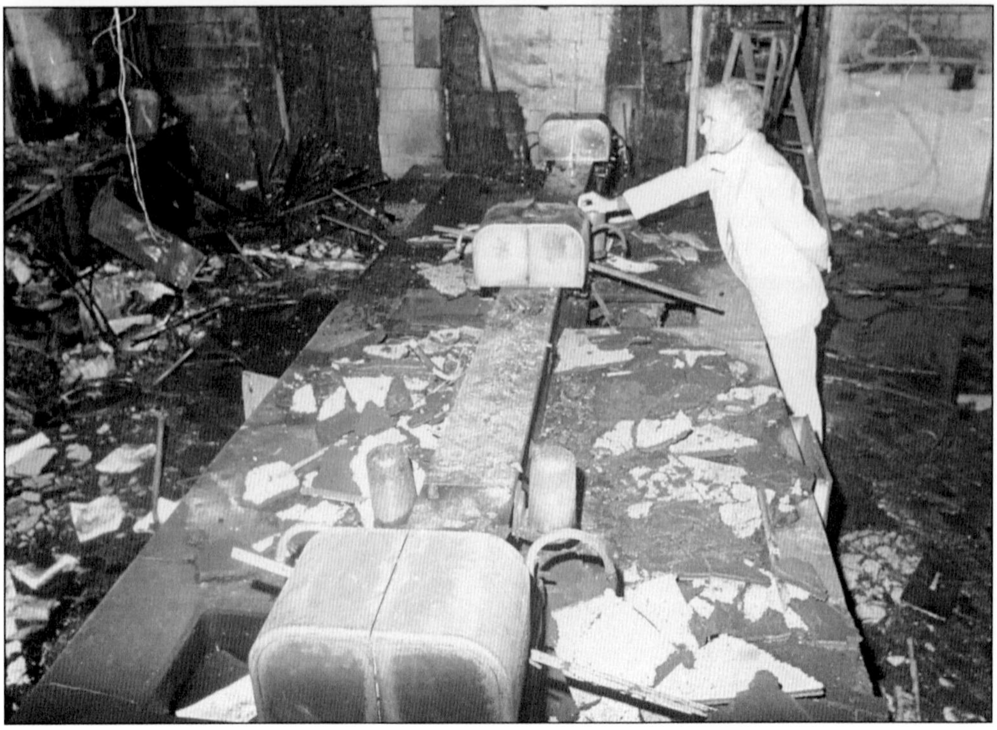

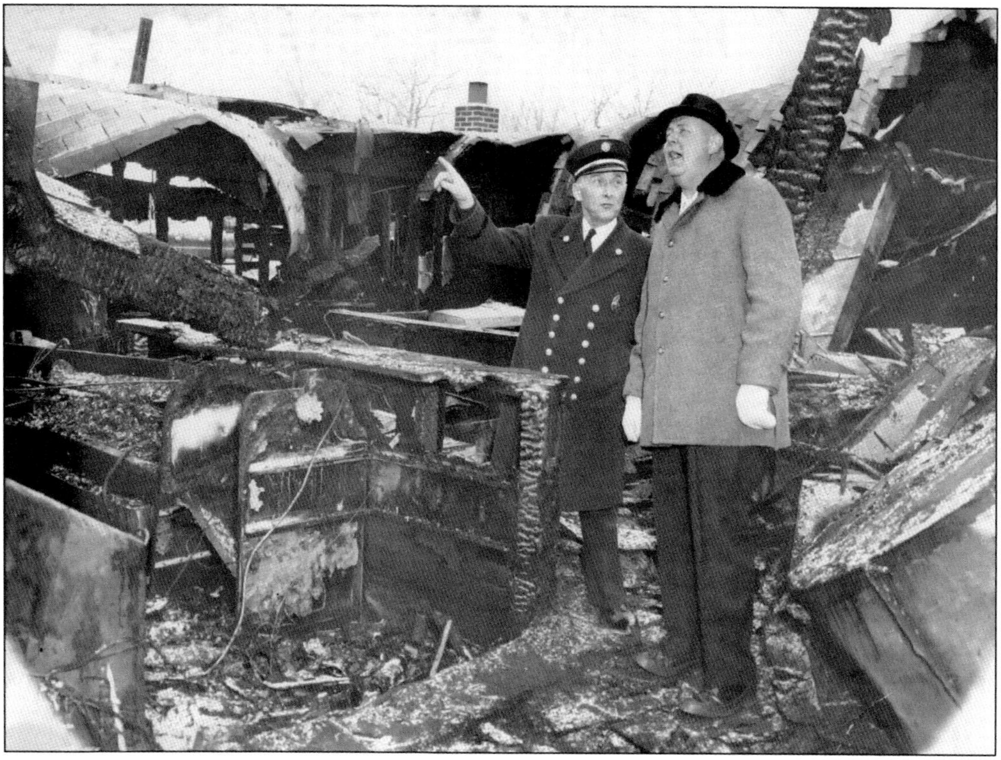

The picture above shows Peabody fire chief Edward P. O'Brine pointing out a location to state police fire inspector Joseph Kelleher while investigating a fire in Peabody. The picture below shows state police fire inspector Sgt. Peter Cummings pointing out a location to fire prevention and arson investigators Capt. Dave Sampson (foreground) and Inspector Joe DiFranco at a more recent fire in Peabody.

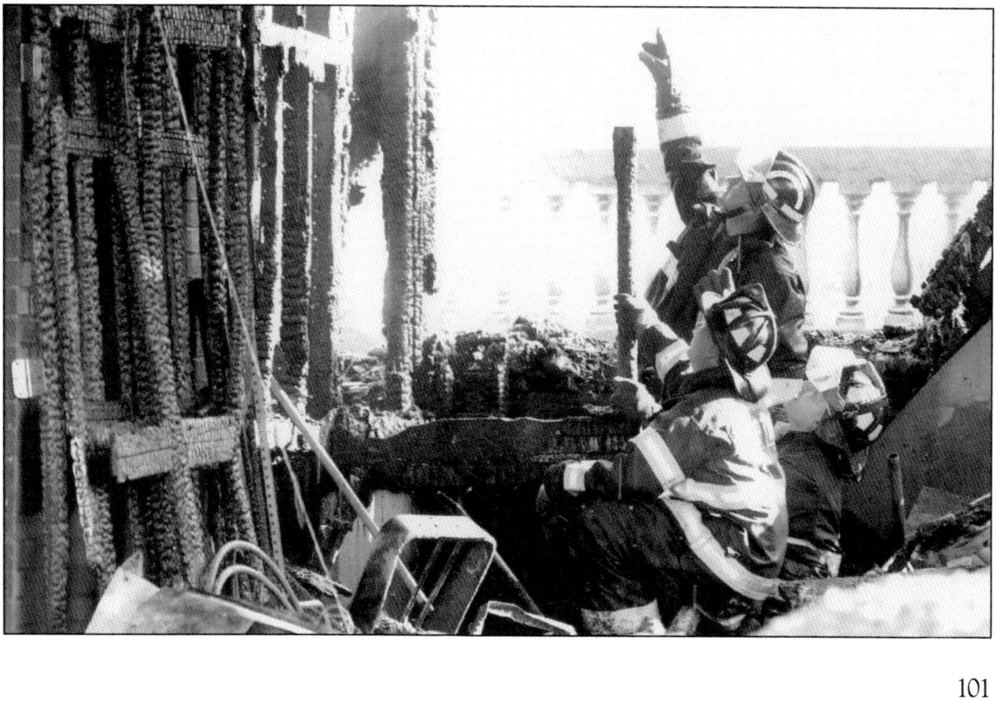

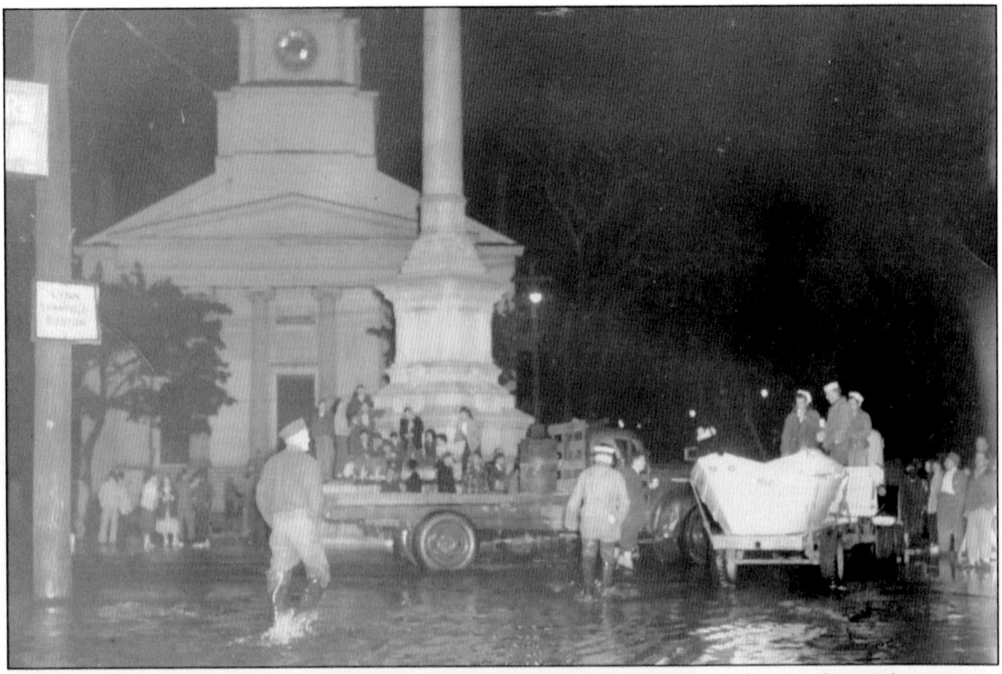

Flooding has always been a problem for Peabody. There were quite a few in the 19th century, and the late 20th century saw them more frequently. The area of Peabody Square was, at one time, a five-acre pond. Called the Mill Pond, it ran down the bank of the north part of Main Street, up Central Street to around Hardy Street, and down Mill Street back up to Main Street. The pond flowed into the North River and out to the ocean a few miles downstream. With all the construction and covering of the brooks with roadways and close proximity to the ocean, flooding is inevitable. The photograph above shows the square when the dam at the flume at Eastman Gelatine let go in 1954, and the photograph below was taken on Spring Street looking toward Foster Street near the square during one of the many large storms that occur every year now.

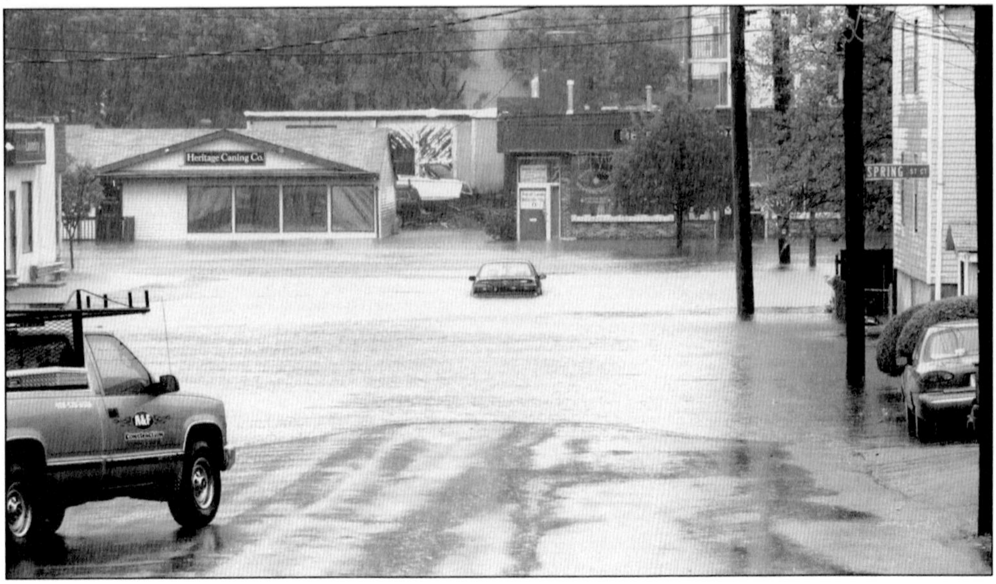

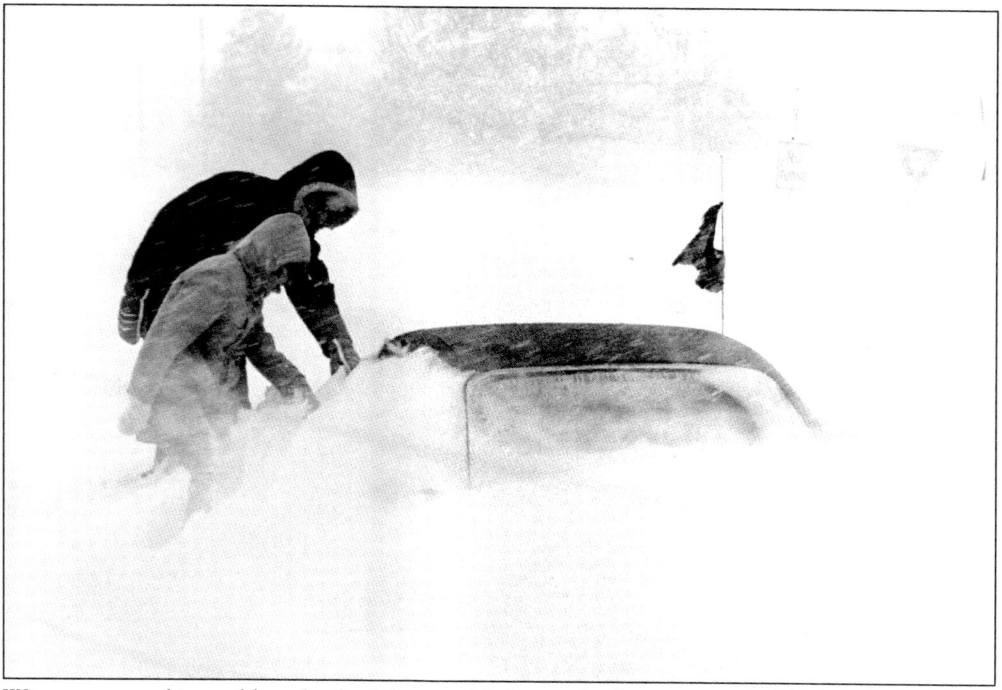

Winter poses a big problem for firefighters with frozen, broken, or covered hydrants, snow- and ice-covered ground, and inaccessible roadways. Some of the biggest fires were during such storms. Pictured here are a couple of photographs from the blizzard of 1978. The picture above shows a couple trying to get to their car that is almost buried in a snowdrift, and the picture below is of the on-ramp to Route 128 south from Route 114 east. The lights in the background are from the North Shore Shopping Center. People were trapped in their cars or had to just abandon them.

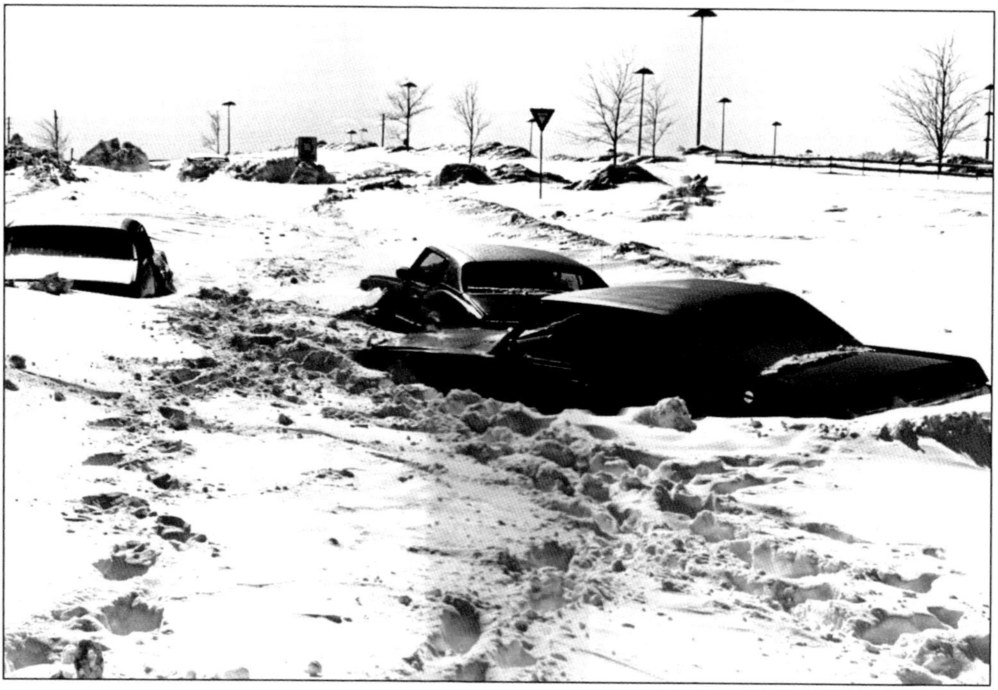

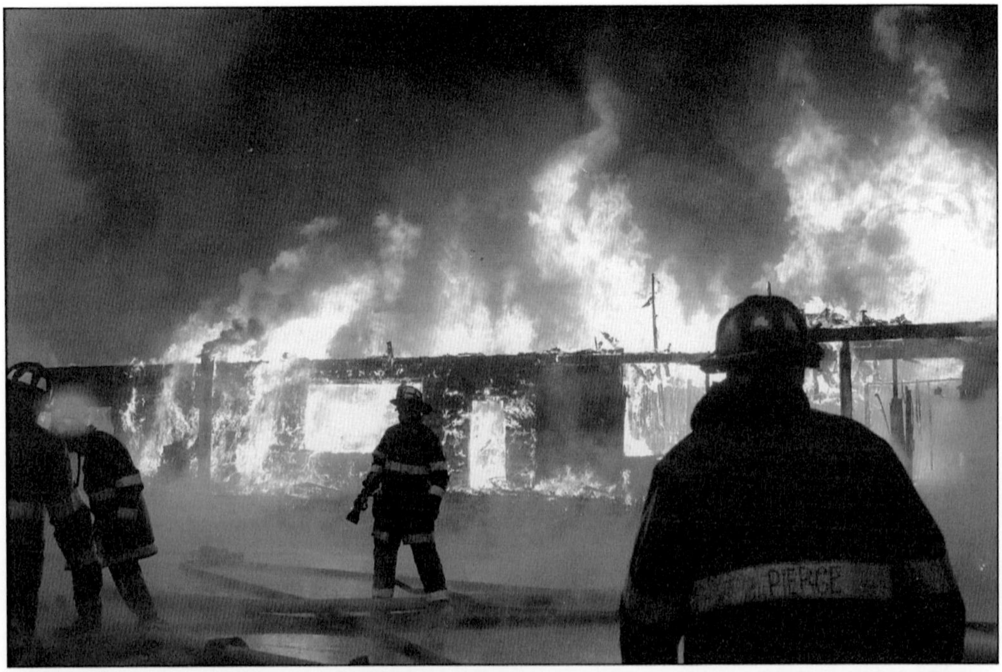

Lowe Mart strip mall on Lowell Street caught fire on the morning of February 27, 1995. It took many hours to get this fire under control. Pictured are Phillip Pierce (right) and Dave Sampson (center).

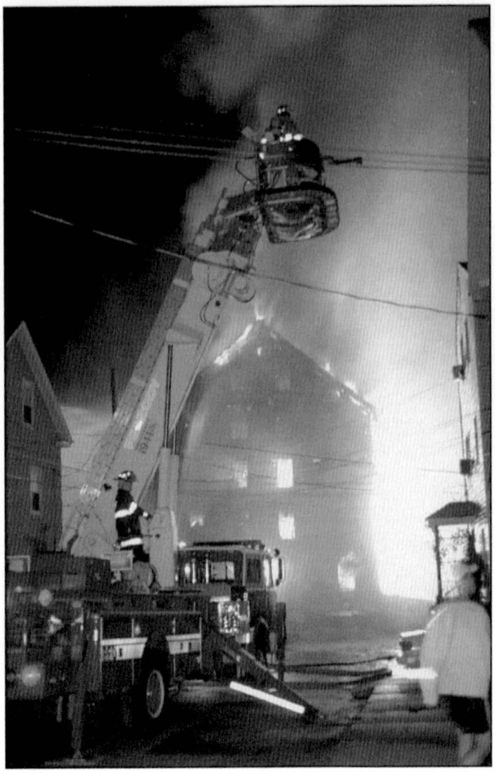

The Bay State Chemical Company building burst into flames around 11 p.m. on May 28, 1996. It was a very large old vacant building on Union Street near Paleologous Street.

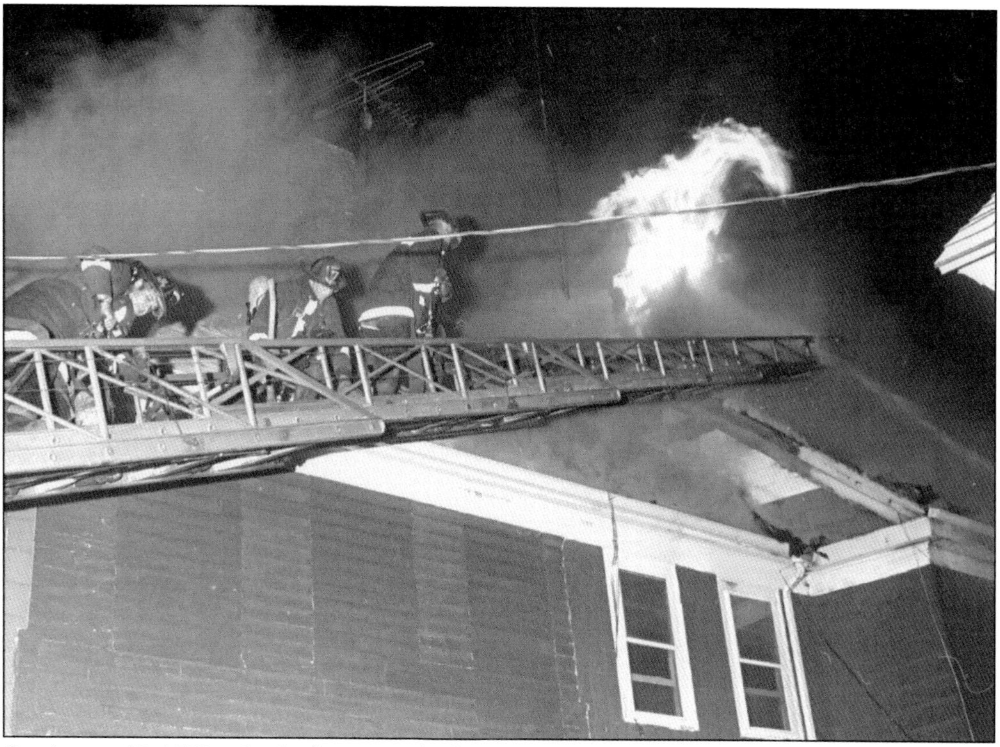

On August 29, 1979, a fire broke out at the home of the Davidson family on Aborn Street.

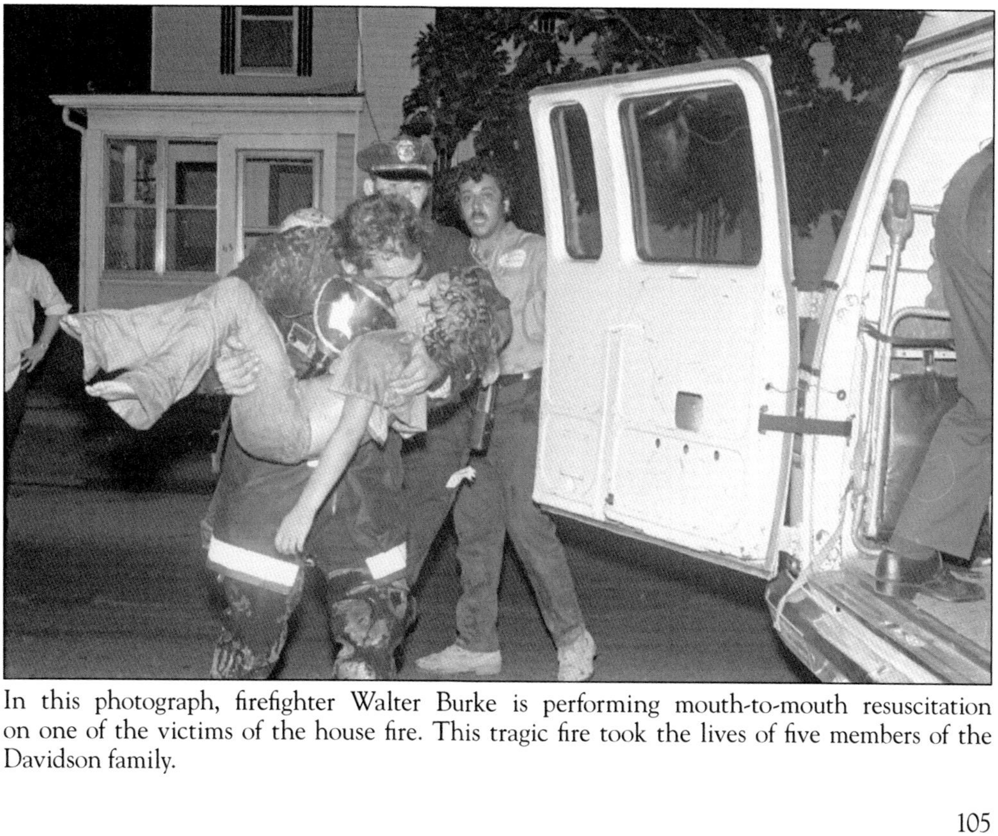

In this photograph, firefighter Walter Burke is performing mouth-to-mouth resuscitation on one of the victims of the house fire. This tragic fire took the lives of five members of the Davidson family.

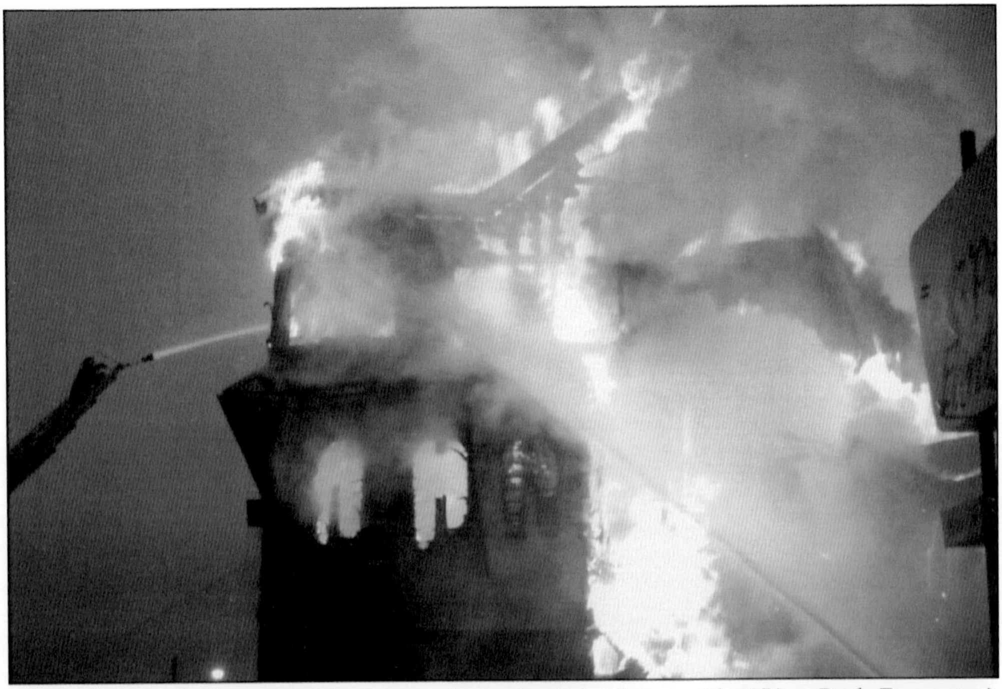

These photographs show an early-morning fire on Saturday, January 13, 1979, at Paula Fountaine's Lounge (the former Palm Garden). Box 542 was struck at 5:59 a.m. Firefighter James Daley went to the Josiah B. Thomas Hospital, and firefighter Ed Tyburc hurt his back. Also on January 13, 1979, there was a house fire at 14 Bartholomew Street with heavy fire and smoke in the cellar.

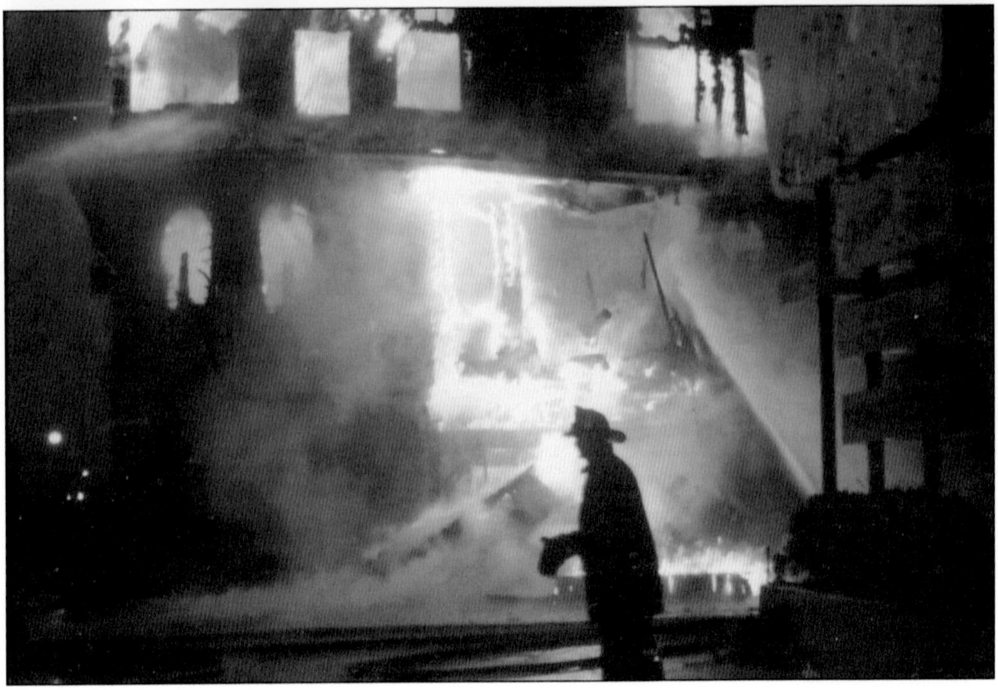

Lt. Joe DiFranco calls from the window for something at a fire at the old tannery of A. C. Lawrence in the Q building.

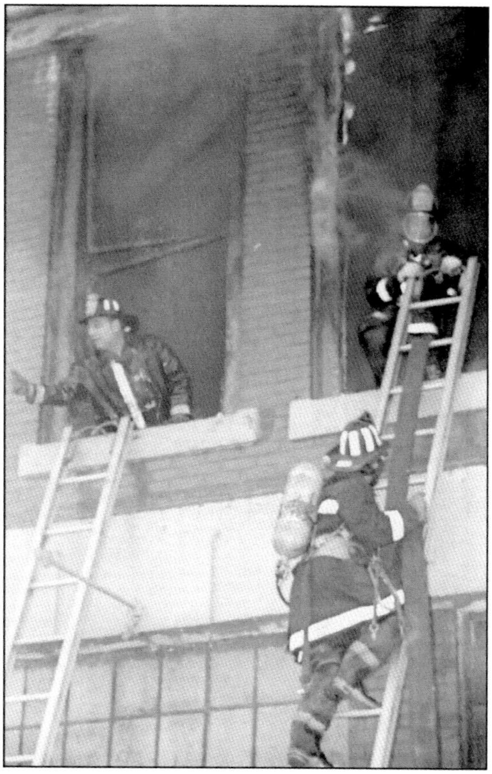

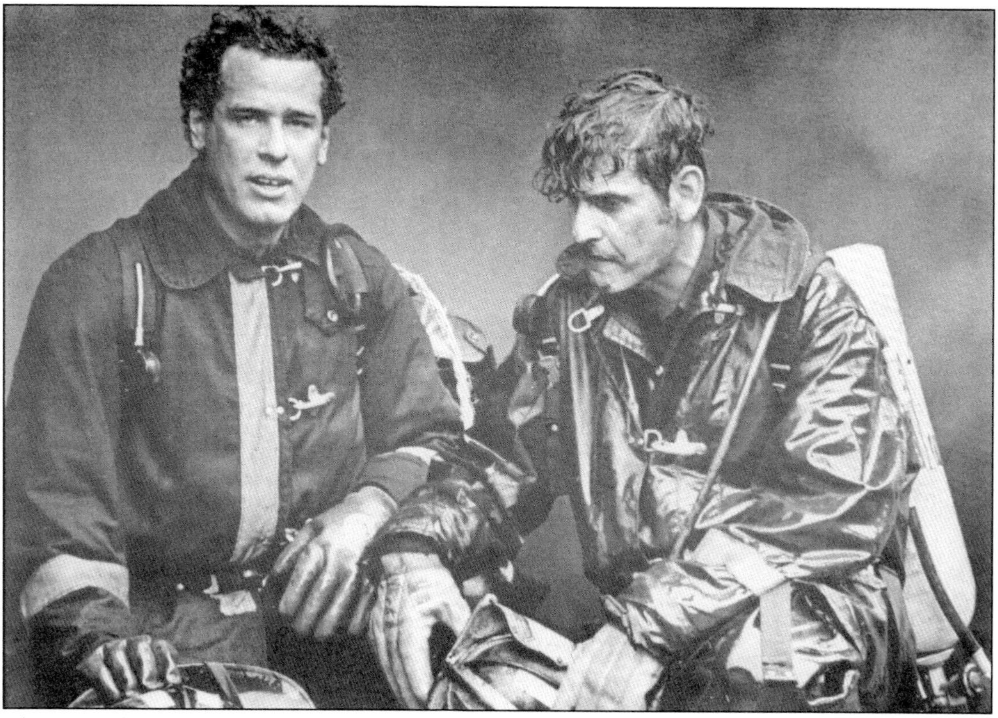

This is a classic photograph of firefighter Bob Caruso (left) and Capt. Sid Durkee taking a breather after a fire at one of the Zolotas buildings on Walnut Street.

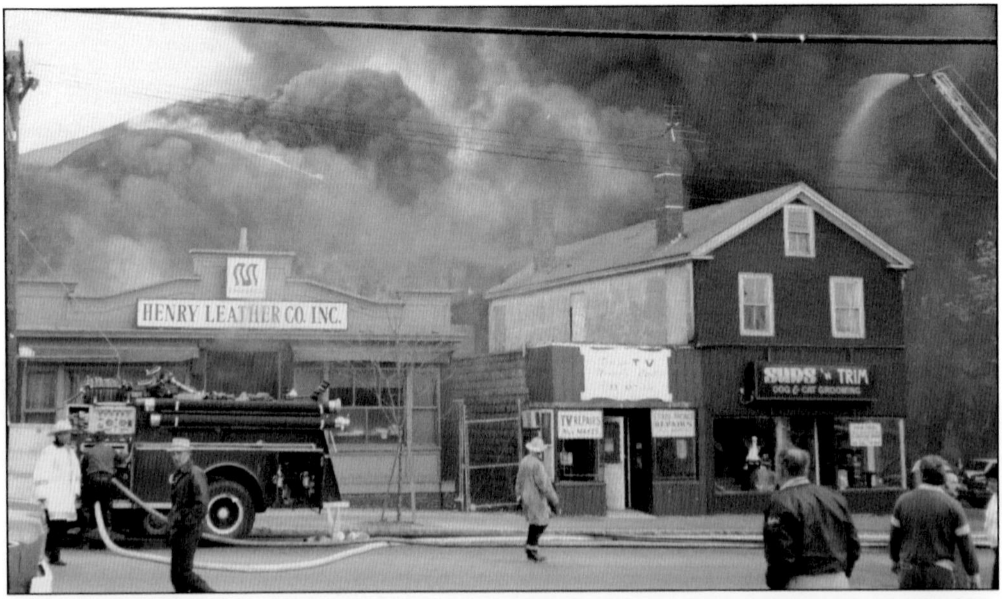

May 10, 1984, started out like any other morning. Hose training was commencing on Perkins Street, and the daily routine of most continued until 9:30 a.m. when an explosion rocked the downtown area of Peabody. On Main Street at the Henry Leather Company, one of the last of the big leather factories burst into flames, threatening the entire area. The conflagration that followed filled the air with thick, black, billowing smoke with flames shooting hundreds of feet in the sky.

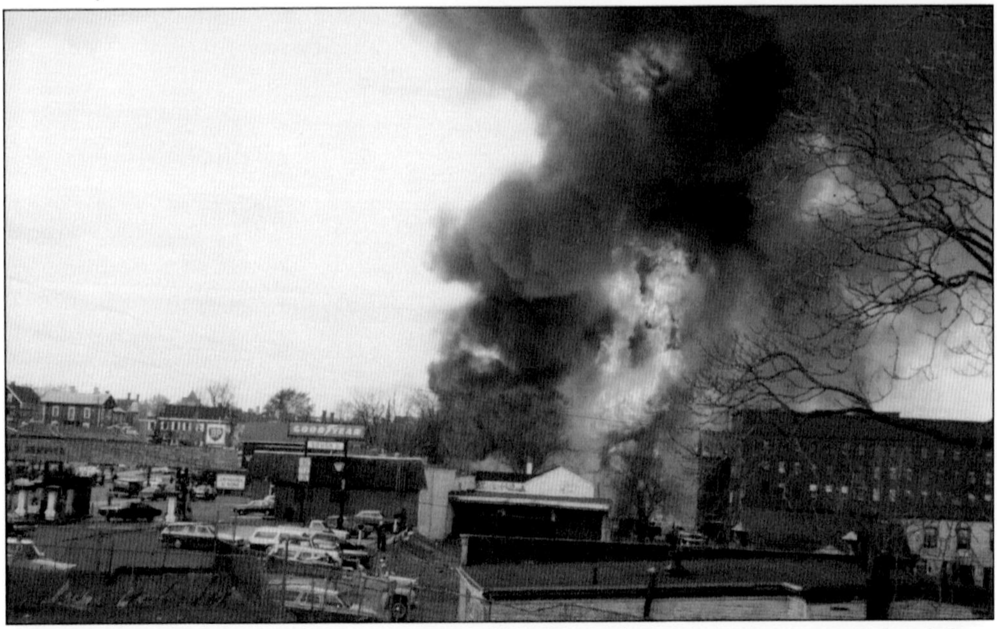

The duty officer that day, Deputy Joe Russell, when referring to getting mutual aid from surrounding communities, was quoted as saying, "Don't stop it from coming and heavy on the ladder truck." His request was received with the attendance of 37 engines and 10 ladder trucks, including airport crash trucks from Logan Airport and Hansom Air Force Base. Assistance reportedly came from as far away as New Hampshire.

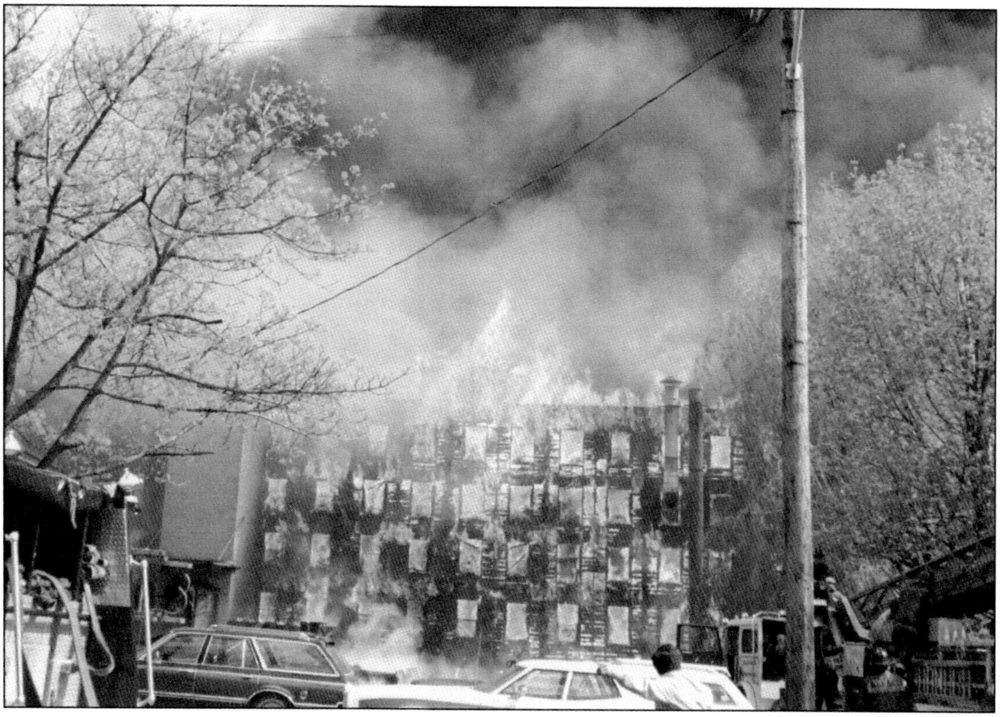

The Henry Leather Company fire is considered to be one of the biggest fires in Peabody history. Called the May 10th fire, it is one the only fires in Peabody that is referred to by its date. Amazingly only one life was lost, a worker on the roof who was cutting a hole. His body was found the next day. The total loss from the fire was in the tens of millions of dollars. Losses from the fire included 19 businesses, 300 jobs, and 27 motor vehicles, as well as having a major environmental impact. It was one of the lead stories in *Firehouse* magazine, a firefighting journal.

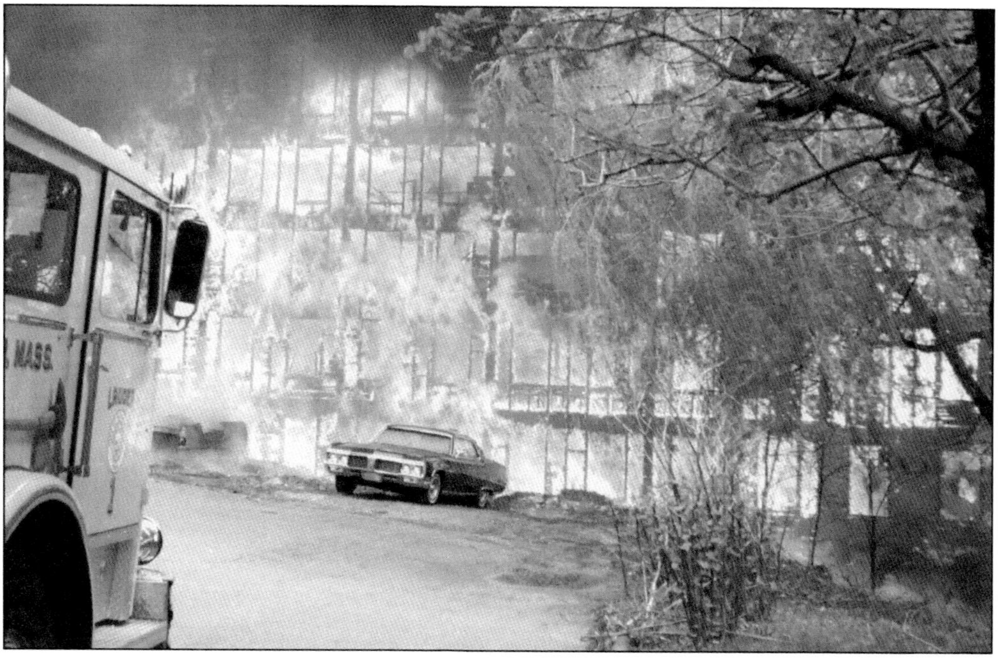

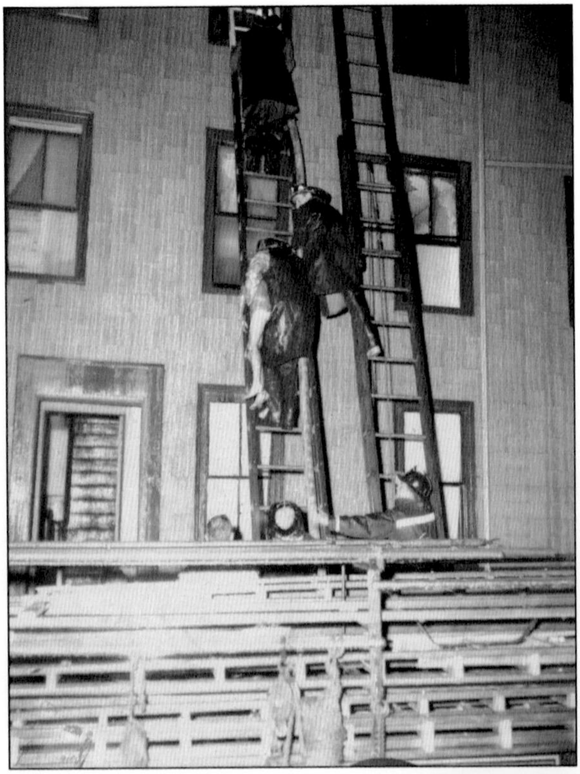

This picture shows a dramatic ladder rescue taking place from a third-floor window at an unknown location in the downtown Peabody area.

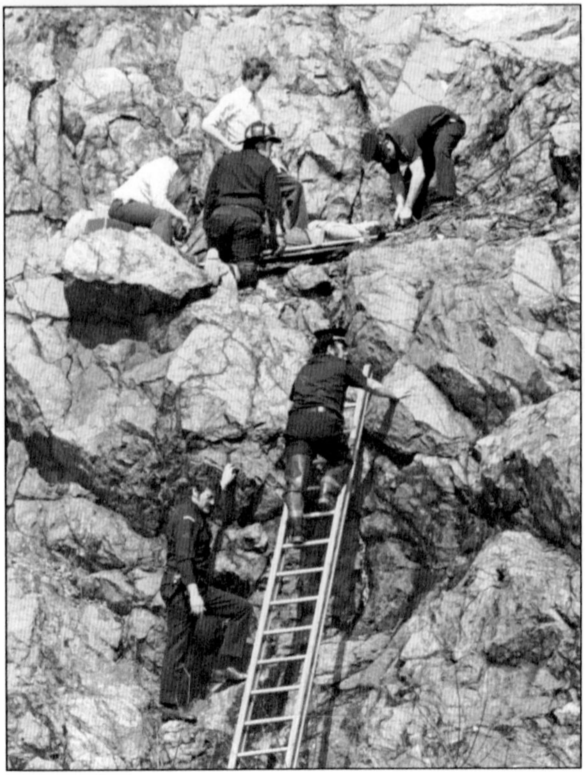

On sunny Saturday, March 20, 1976, a 14-year-old boy fell approximately 40 feet and was trapped on the rocks at the old quarry (Essex Trap Rock), commonly known as Buxton Cliff, on Warren Street Extension. He was not injured except for some bruises on his back. Had he not landed in that spot, he may have fallen another 50 feet or possibly to his death. Police officer Larry Grant and firefighters Kenny Burkinshaw, Lenny Marshall, Phil Tighe, Ed Tyburc, Stanley Smith, Capt. Tony Koutzoukis, Lt. Henry Velez, and acting deputy Jim Carter made the rescue.

Six

THE FACES AND NAMES

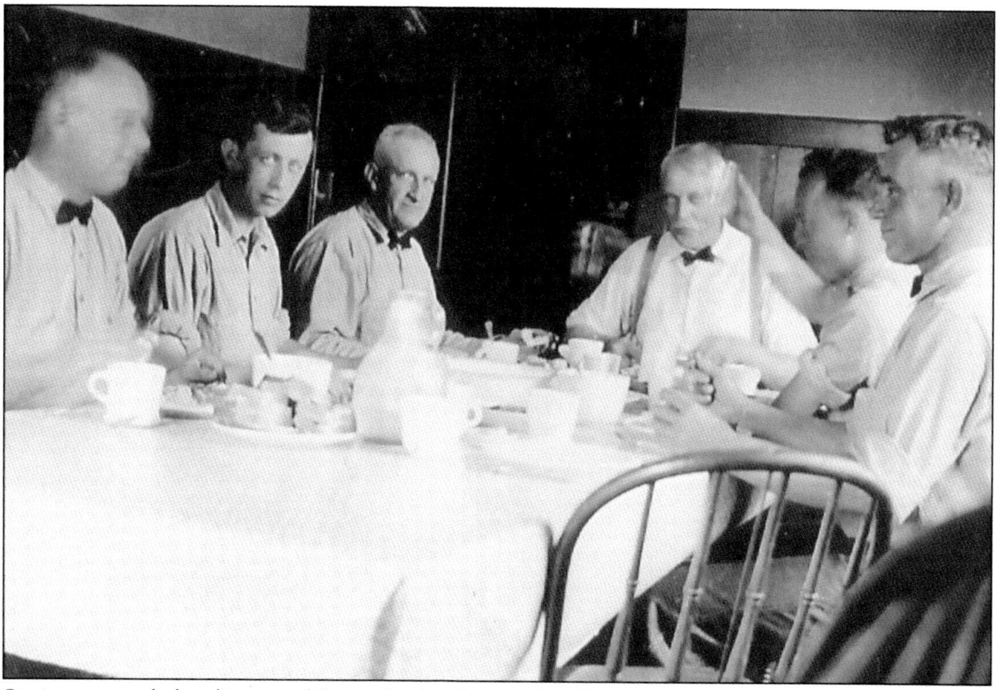

Sitting around the dinner table in the kitchen at headquarters are, from left to right, Joseph Mulcahy, William Frame, Cornelius J. Duggan, William Brown, Alfred Coombs, and Raymond Skinner.

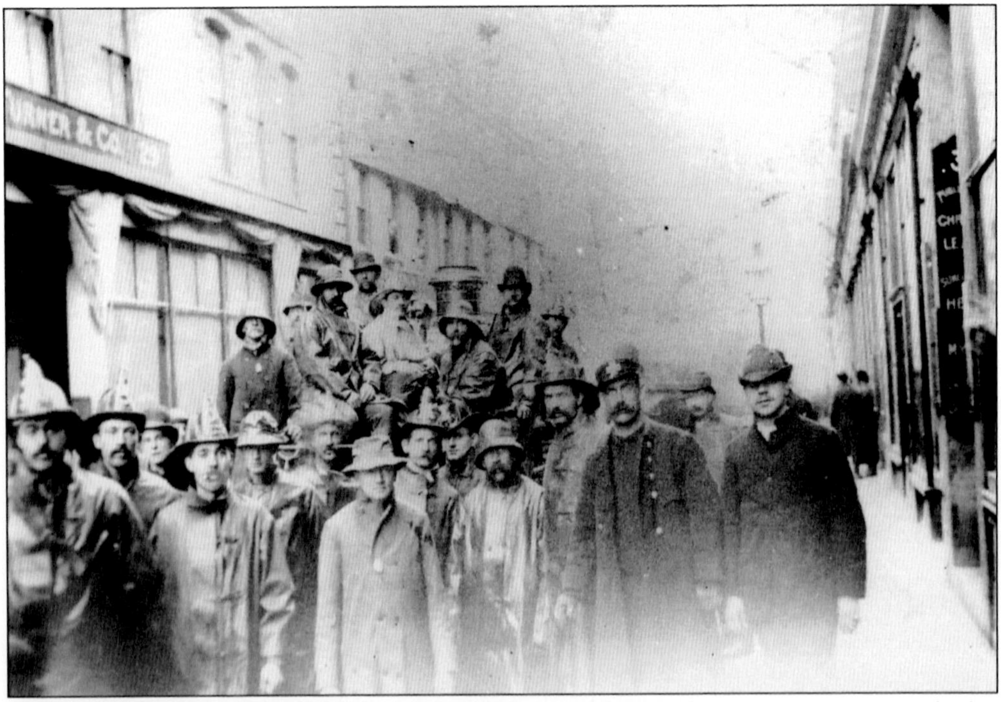

Firemen have always loved posing in front of their apparatus. Pictured above are Peabody's bravest with their steamer engine on a street in either Lynn or Boston.

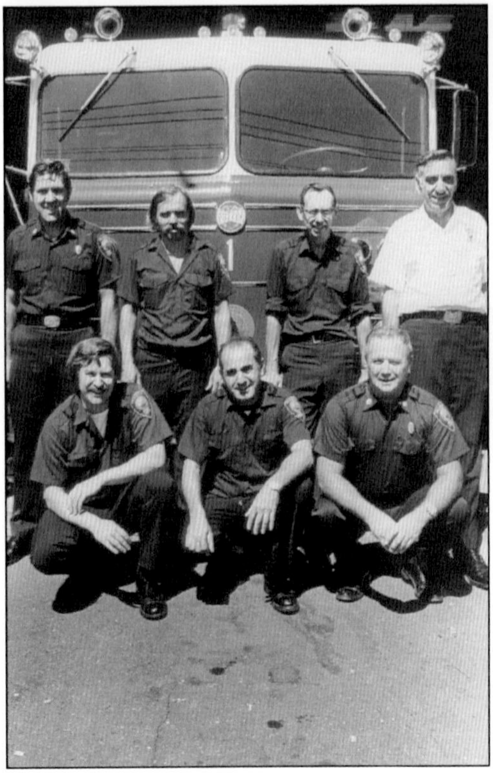

Pictured from left to right in front of engine No. 1 are (first row) Bob Derloo, Ned Peluso, and Don McCauley; (second row) Fred Trueira, Bob McClory, Ray Brown, and Deputy Peter Pappas.

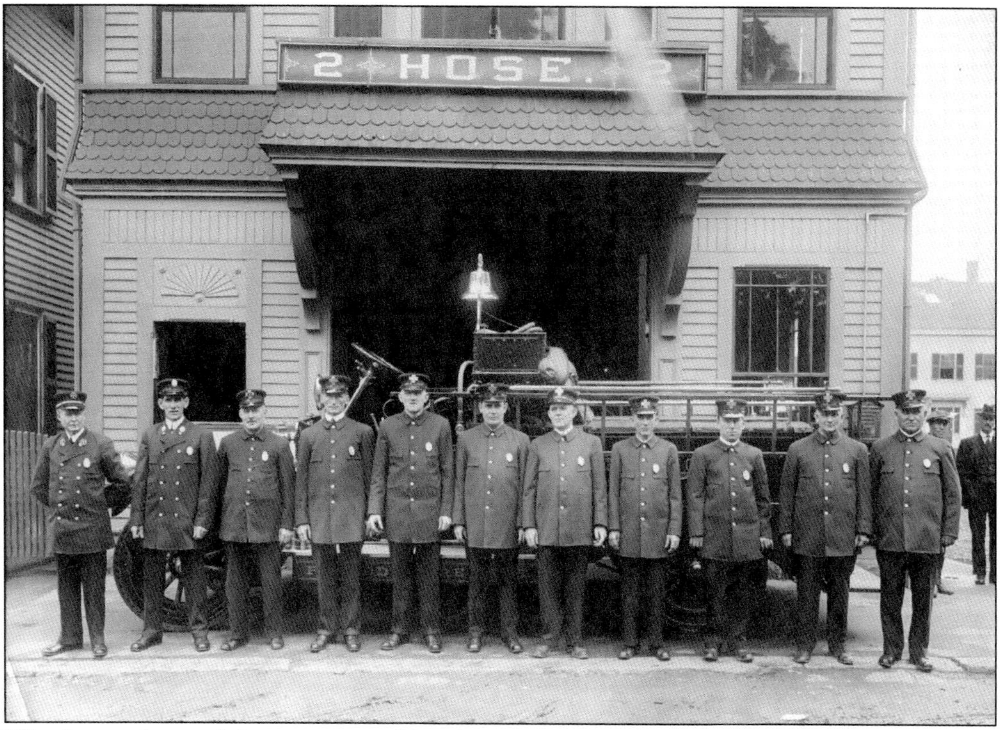

Whether in front of their apparatus like the men in this photograph of engine No. 2 or just hanging out on the "back step" of engine No. 1 at Central Station, the men are always ready to pose with their apparatus. Below, from left to right, are Ray Nelson, unidentified, Charles Kenealy, Ralph Innis, Lenny Lucey, Peter Pappas, and William Frame.

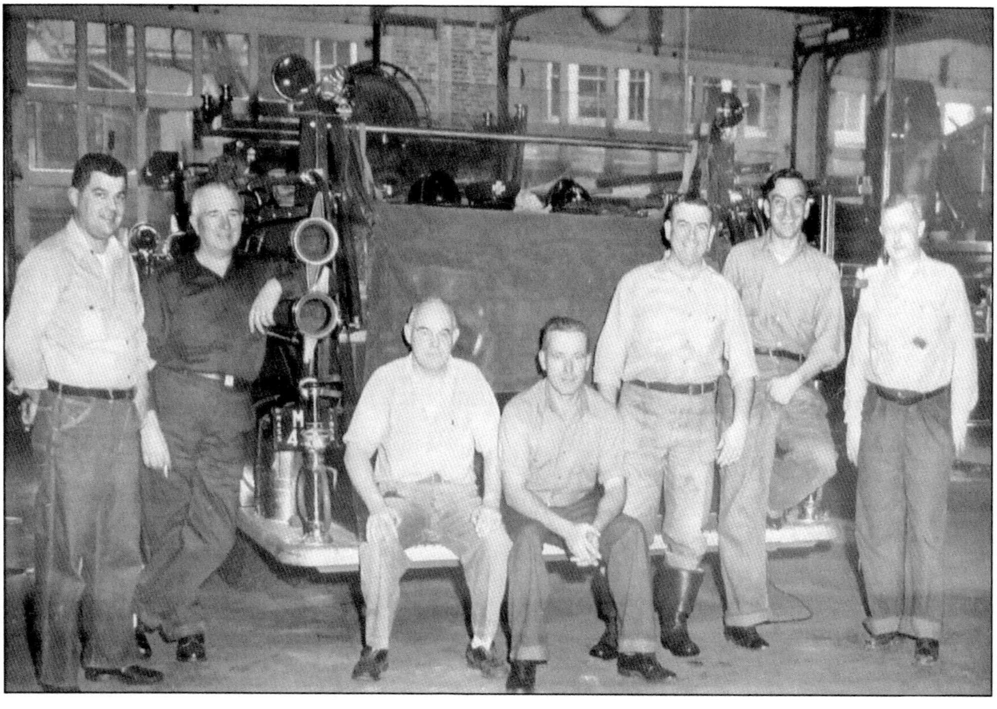

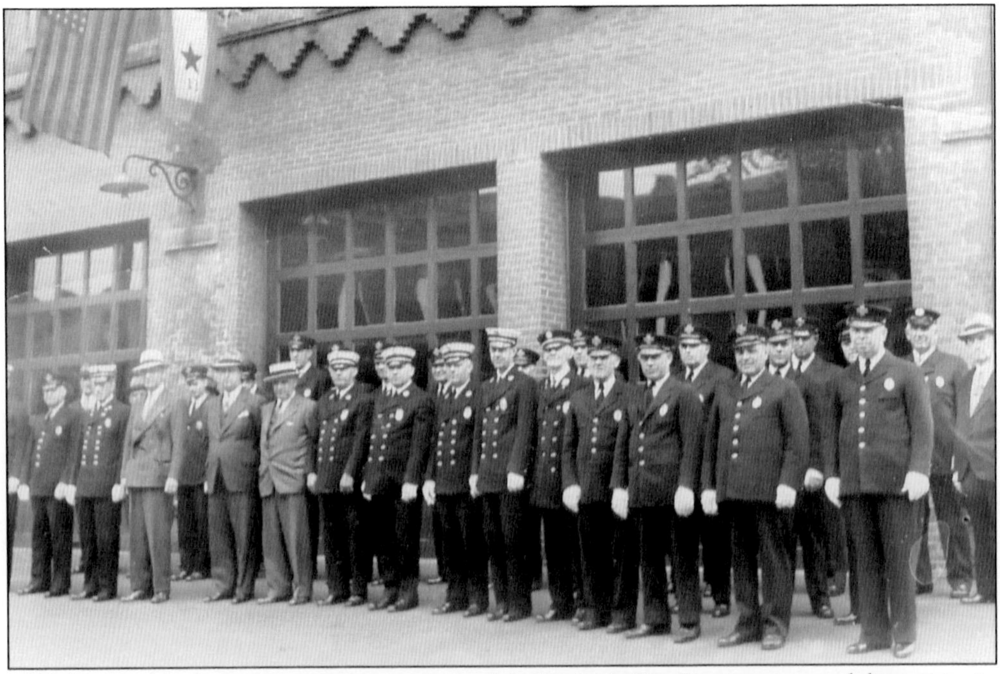

The photograph above shows some members of the Peabody Fire Department and dignitaries in front of Central Station during World War II. Notice the American flag hanging and the service flag with the star and the number 13 indicating the number of men who were serving in the military. The photograph below shows some members of the Peabody Fire Department in front of Central Station just after the renovations in 1988.

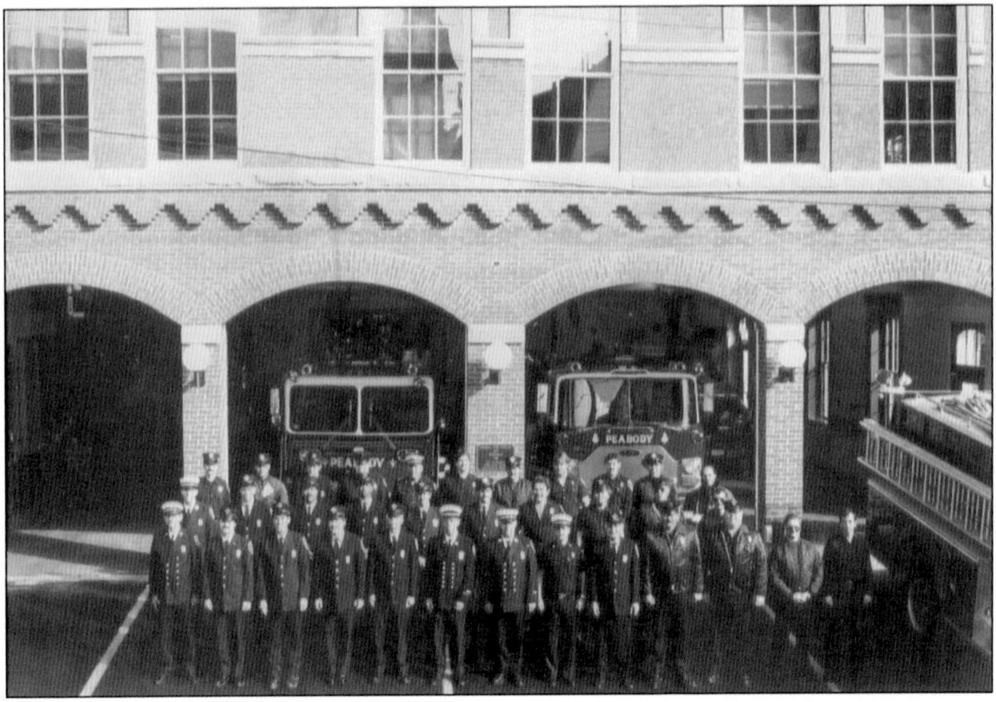

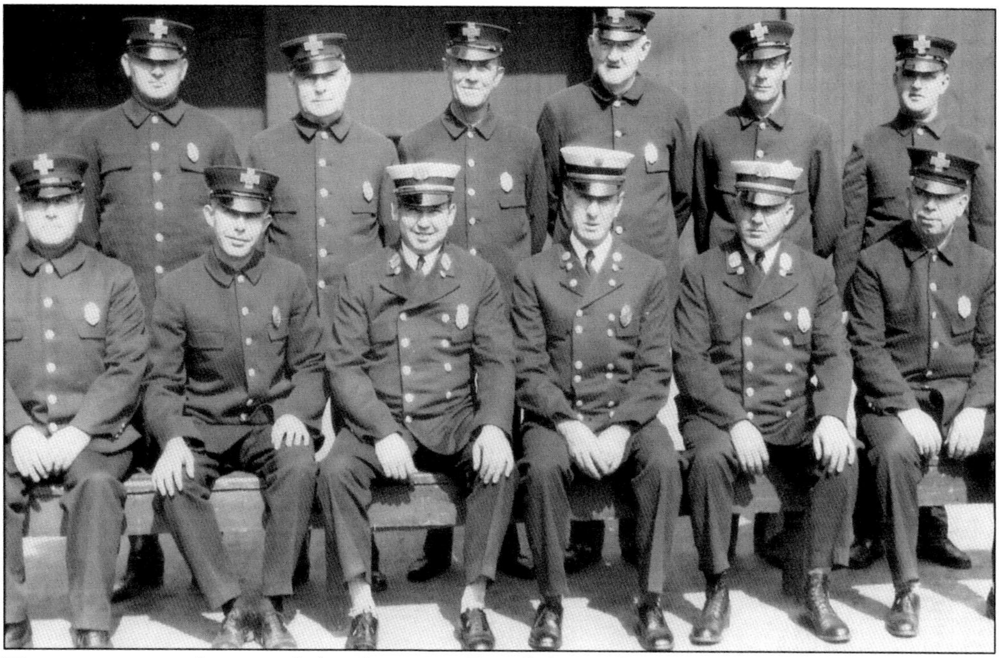

This picture was taken outside headquarters. From left to right are (first row) unidentified, Francis Driscoll, Capt. Charles Lawrence, Chief William Mahoney, Capt. James Carr, and Timothy Foley; (second row) unidentified, Fred Sherry, Charles Fox, Thomas Walsh, unidentified, and George Jacques.

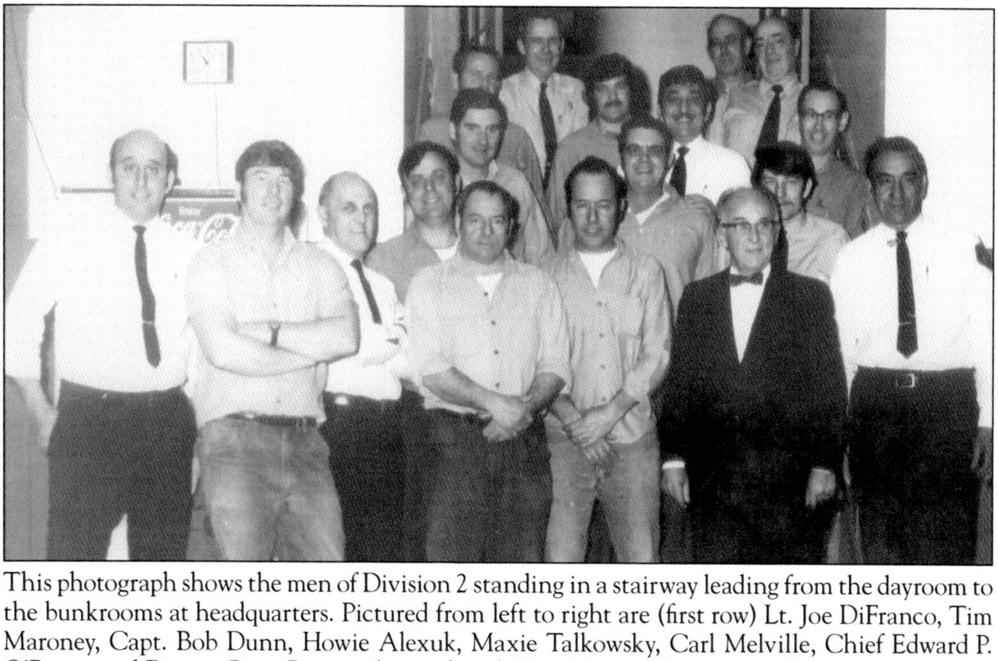

This photograph shows the men of Division 2 standing in a stairway leading from the dayroom to the bunkrooms at headquarters. Pictured from left to right are (first row) Lt. Joe DiFranco, Tim Maroney, Capt. Bob Dunn, Howie Alexuk, Maxie Talkowsky, Carl Melville, Chief Edward P. O'Brine, and Deputy Peter Pappas; (second row) Tom Moran, George Carafa, and Peter Fallon; (third row) Ray Smith, Ed Blanchard, Capt. Tony Koutzoukis, and Ray Brown; (fourth row) John Croughwell, Tony Coito, and Harvey Chase. This picture was taken on February 28, 1972, by Dan Rosa, one of the fire department photographers.

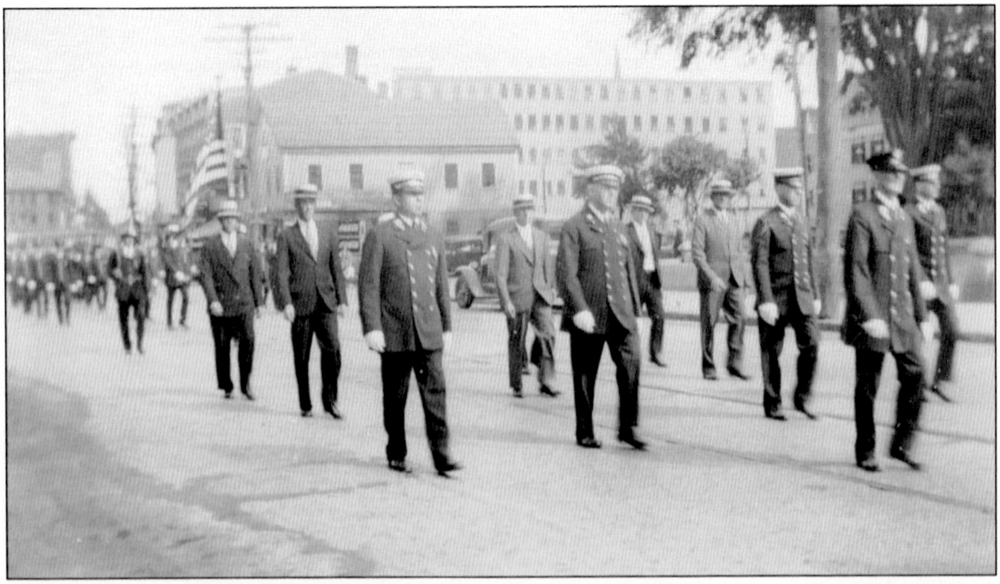

Firemen are seen here parading down the street with city councillors and marching past city hall in one of many Memorial Day parades that firemen take part in.

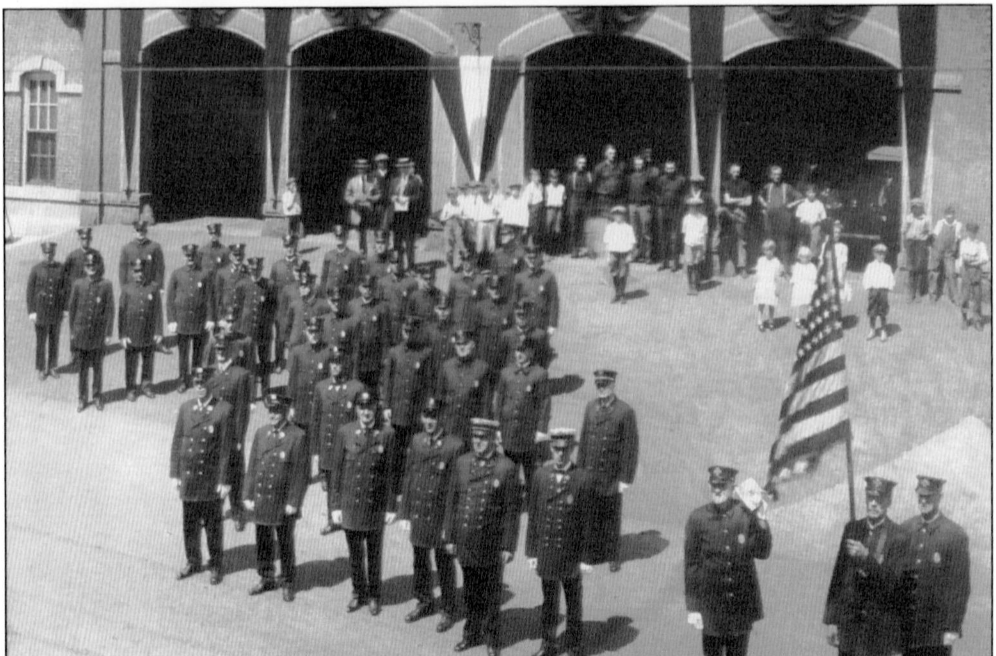

This old photograph, showing men in front of Central Station, might have been taken for the funeral of Cornelius J. Duggan in 1933. Notice the funeral bunting on the bricks.

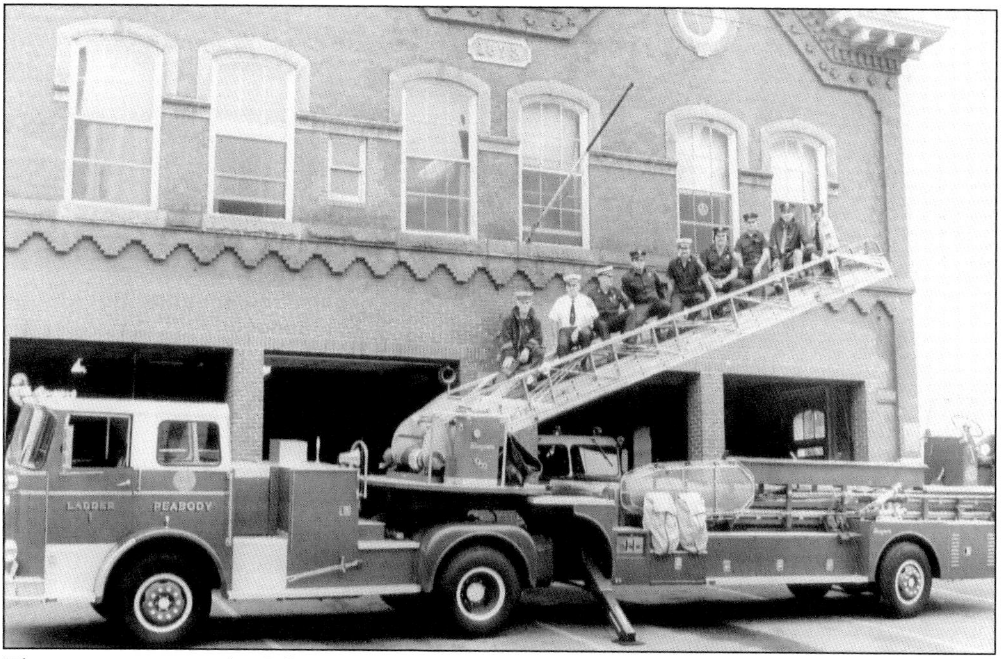

This picture was in the *Salem Evening News* and with an article about family members being on the fire department. From left to right, up the ladder, are Lt. James Kerwin, Deputy Frank Connors, Capt. Eugene Connors, firefighters John Connors, Bill Downey, Paul Lynch, Stan Smith, Ned Peluso, and Bob McClory.

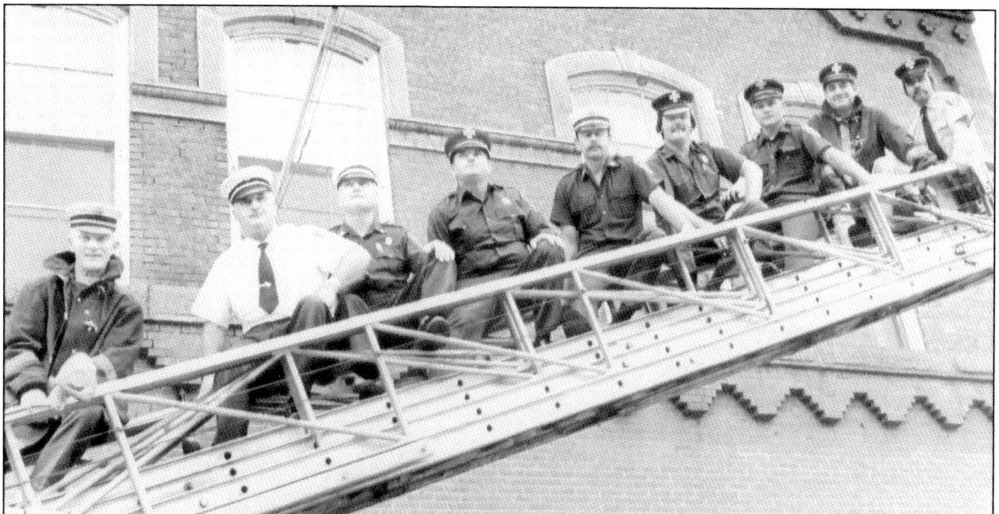

Lt. James Kerwin's brother, John, was on the fire department with him. Deputy Frank Connors, Capt. Eugene Connors, and firefighter John Connors were brothers. Bill Downey had two brothers-in-law on the job with him. Paul Lynch had, on the department, his brother, Kevin, who died in the line of duty in 2001. Stan Smith's twin sons, Eric and Scott, were appointed to the fire department just prior to Smith's retirement. Smith's brother Brad was on the Peabody fire department as well. Ned Peluso's son, Mike, was appointed to the fire department in Peabody just prior to Peluso's retirement. These are just some of the names of the fathers, brothers, and sons of the Peabody Fire Department.

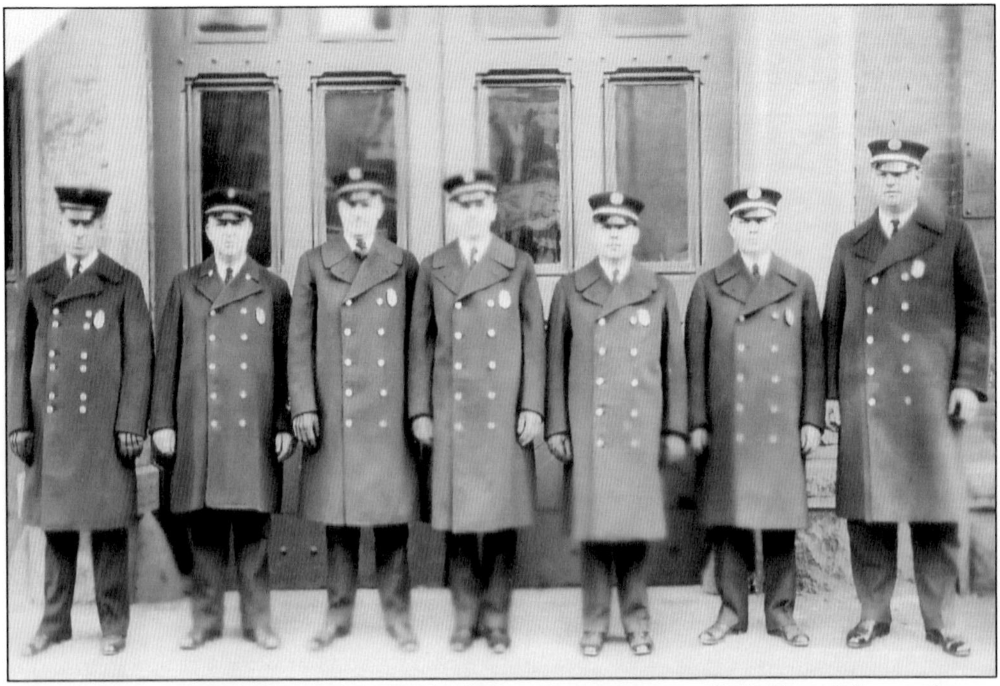

This picture is of the officers of the Peabody Fire Department in 1933. From left to right are Chief William Mahoney, Deputy Daniel Hannon, Capt. Arthur Flynn, Capt. William Costello, Capt. Charles Lawrence, Capt. James Carr, and motor supervisor Joseph Sweeney.

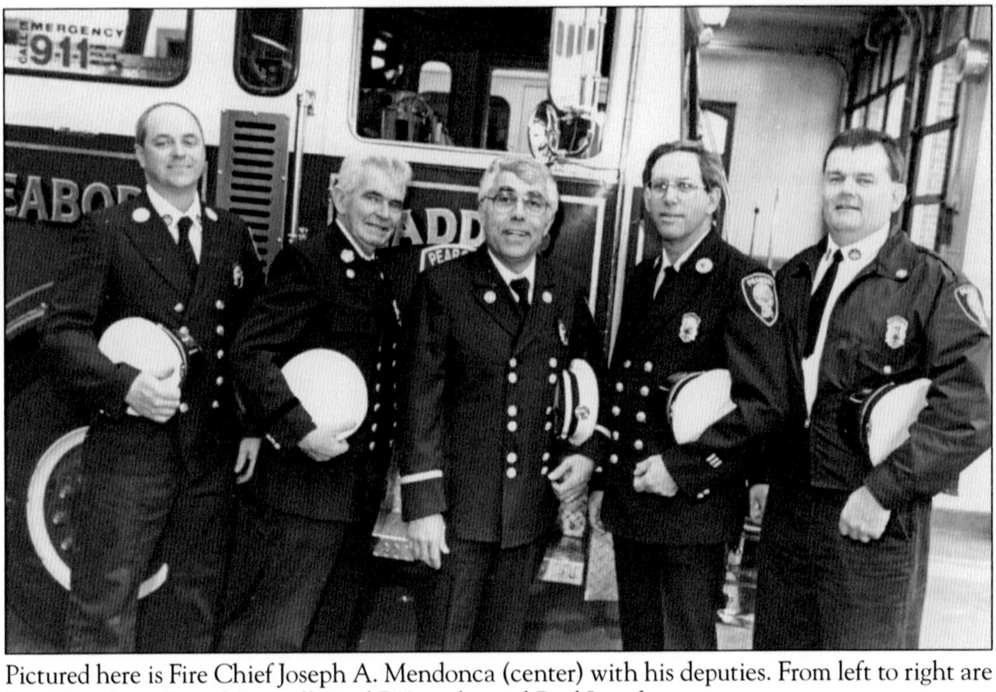

Pictured here is Fire Chief Joseph A. Mendonca (center) with his deputies. From left to right are Paul Hinchion, Joseph Russell, Paul D'Angelo, and Paul Lynch.

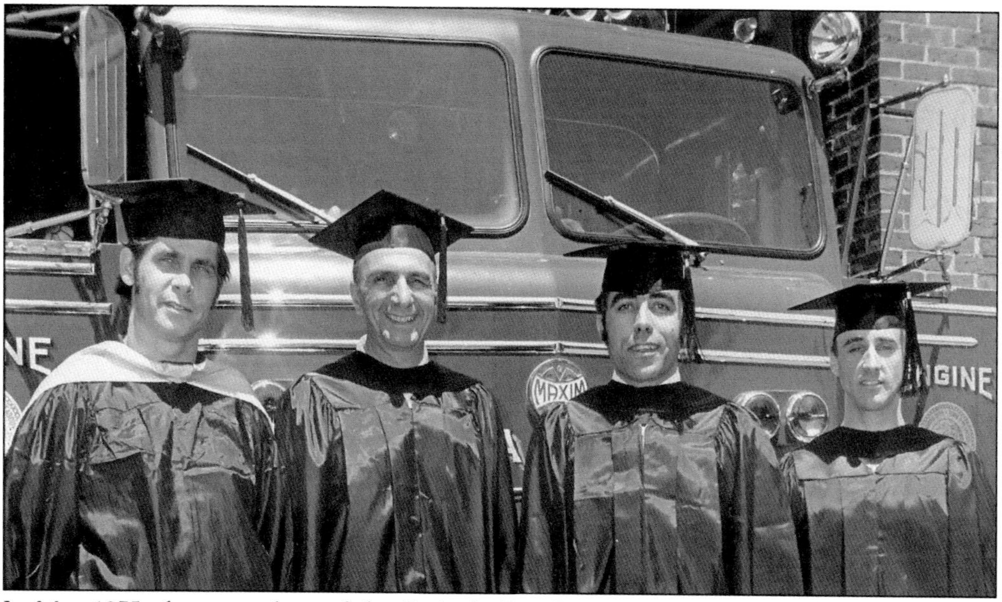

In May 1975, three members of the Peabody Fire Department graduated from North Shore Community College. Frank Ryan (far left) was their instructor. The graduates are, from left to right, Sal Mistretta, Joseph Mendonca, and Mark Phillips.

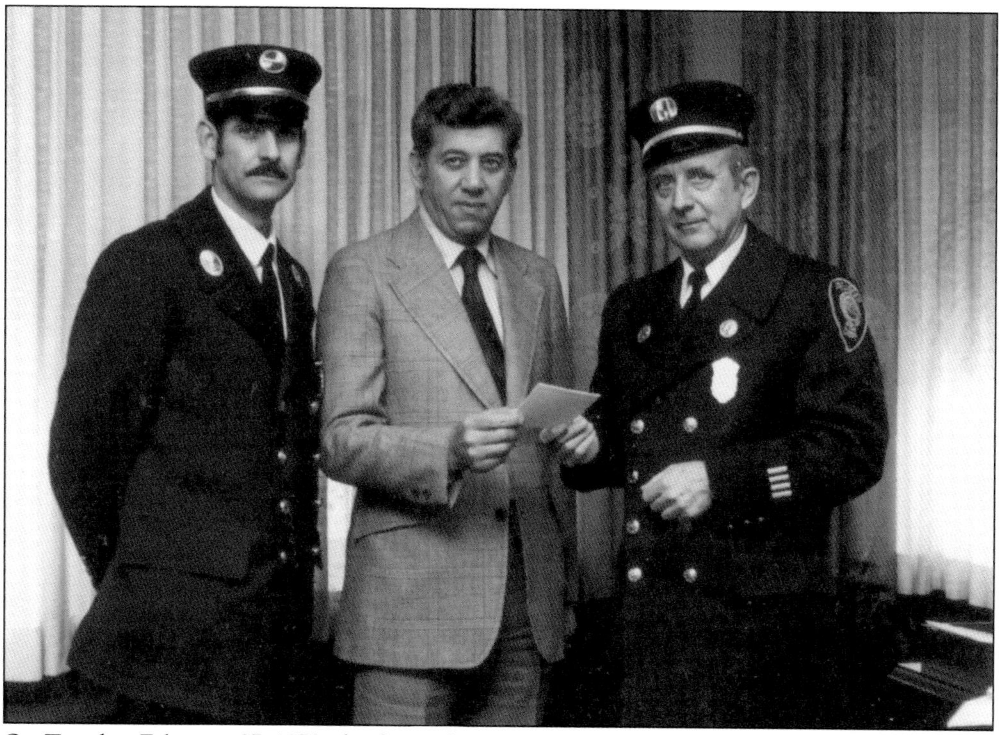

On Tuesday, February 27, 1973, the first ticket sold for the John P. Quinn Jr. Scholarship Dance was to Mayor Nichulas Mavroulas (center). John, who was killed while serving his country in Vietnam, was the son of Peabody fire captain Jack Quinn (far right). Capt. Sid Durkee, a lieutenant at the time the photograph was taken, is at far left.

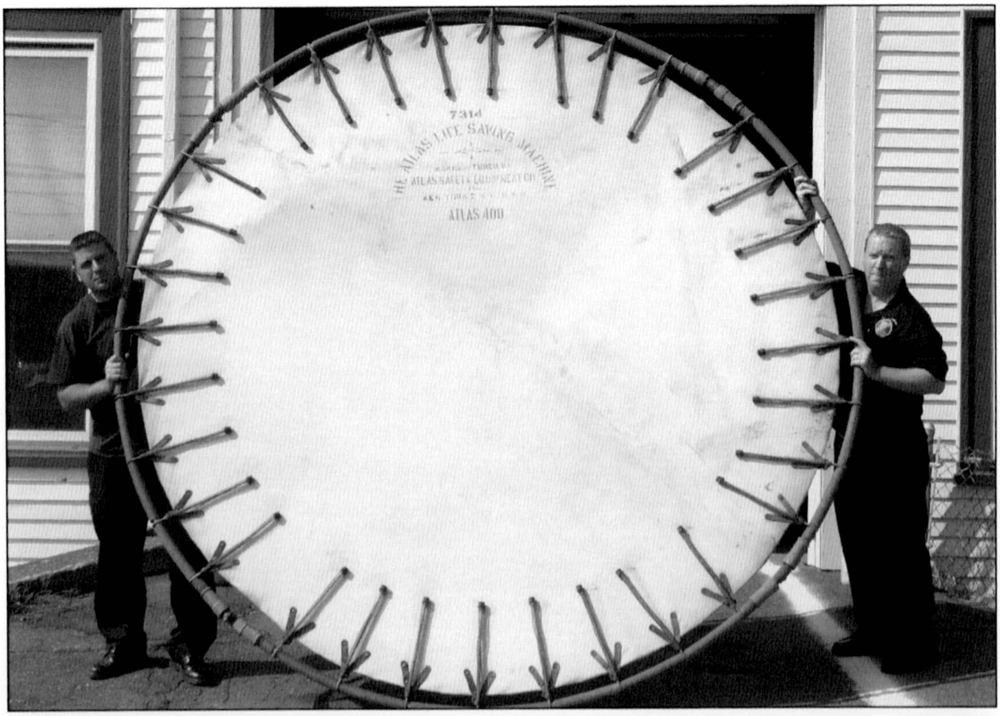

Firefighters Tom Desmond (right) and Tony Zuppio hold a life net that is no longer used. They are standing in front of Engine No. 4 on Tremont Street.

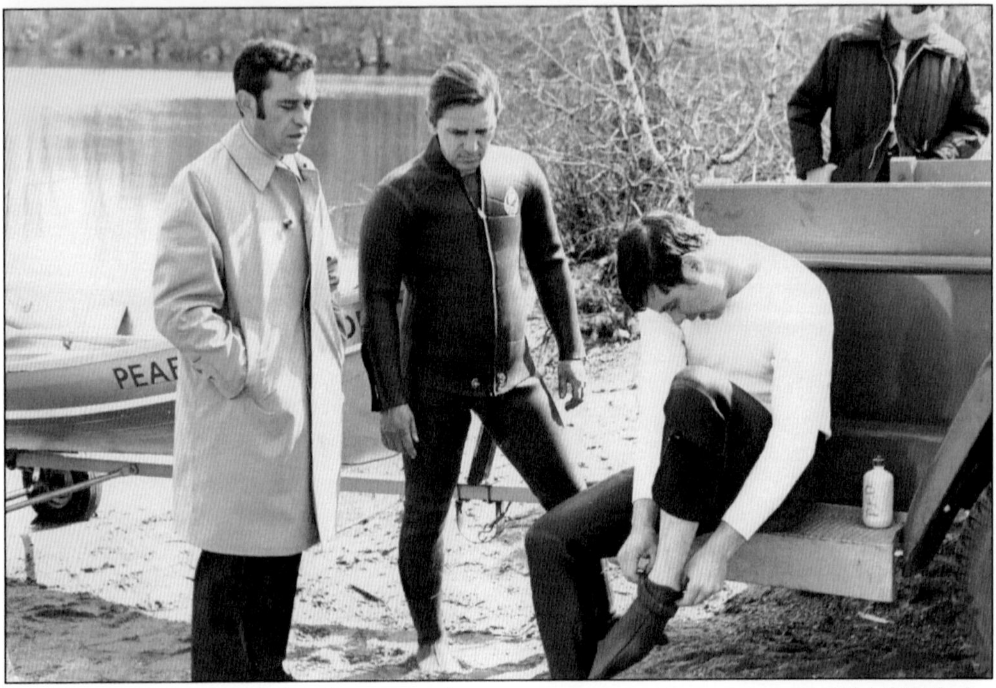

The shore of Brown's Pond is the scene of a training exercise in water rescue. Chief Nicholas Gerakaris (right) stands next to Arthur Blake, while firefighter Tom Davis pulls out his scuba booties. Ray Wallman, a news photographer, is looking on in the background.

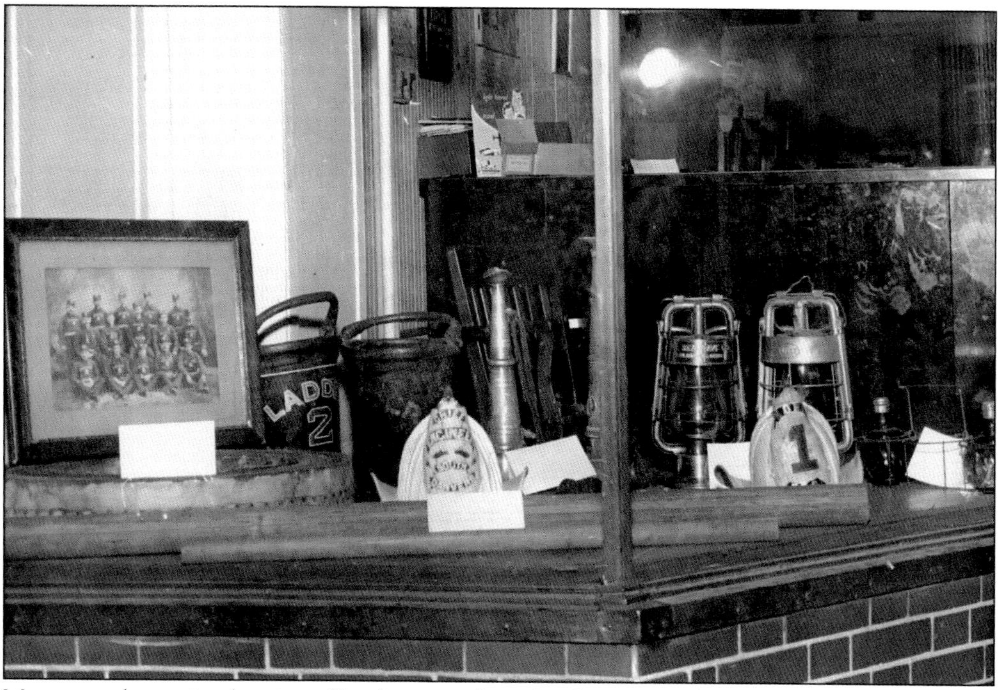

Many merchants in the city offer their windows for decorating during festival weeks such as Fire Prevention Week. These two window displays are from P. J. Woods's insurance company on Main Street. Notice the pictures in the windows; some of them are in this book.

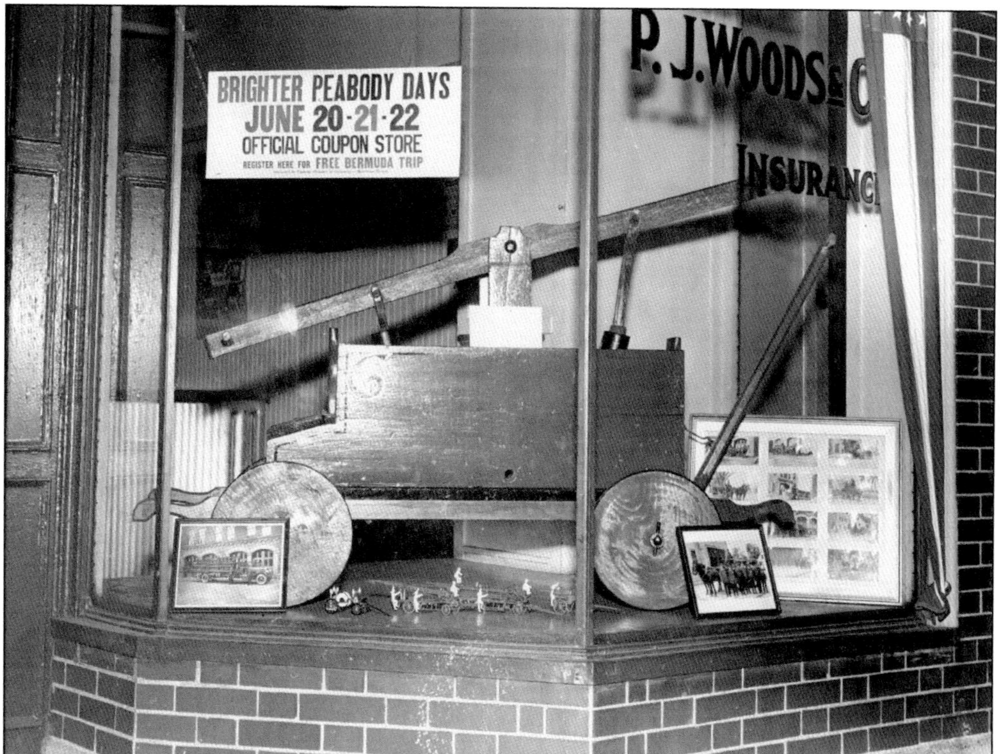

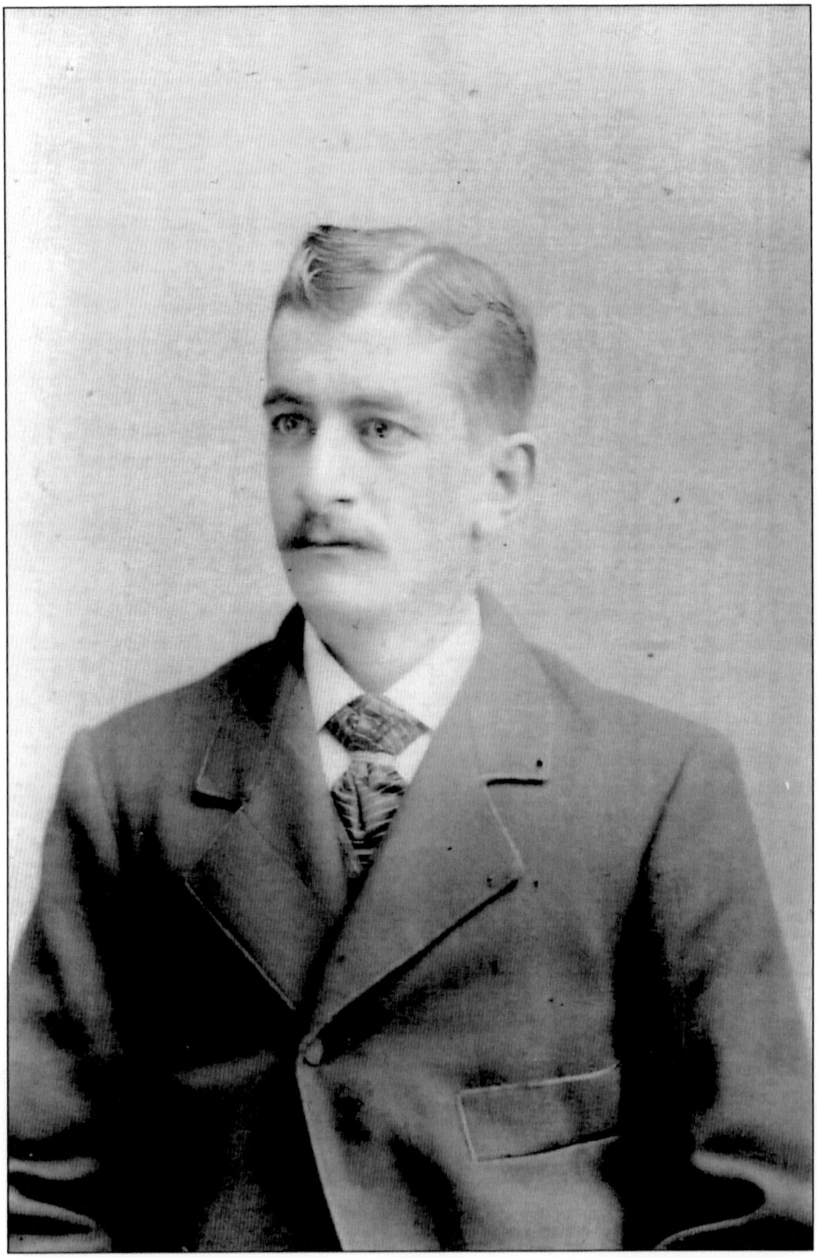

The year 1923 was the last for the old board of fire engineers and its chiefs. The board ran the fire department with the chief engineer as its head. Jesse Fremont Barrett was the last of the old chief engineers due to the newly created position of full-time chief of the fire department. Along with being chief engineer, he also was the superintendent of the water department. Born on July 11, 1856, in South Danvers (Peabody), Barrett, the son of a farmer, married Martha A. Lord. He raised his family at 4 Andover Street. He was appointed a call man in 1874 and became a member of the board of engineers in 1901. He was later elected chief engineer and held the position for 17 years until his retirement from the department in 1923. He might be the longest-active firefighter in Peabody with over 49 years on the job. He died on July 13, 1929, at Russell's Grove in West Peabody and was buried in Cedar Grove Cemetery in Peabody.

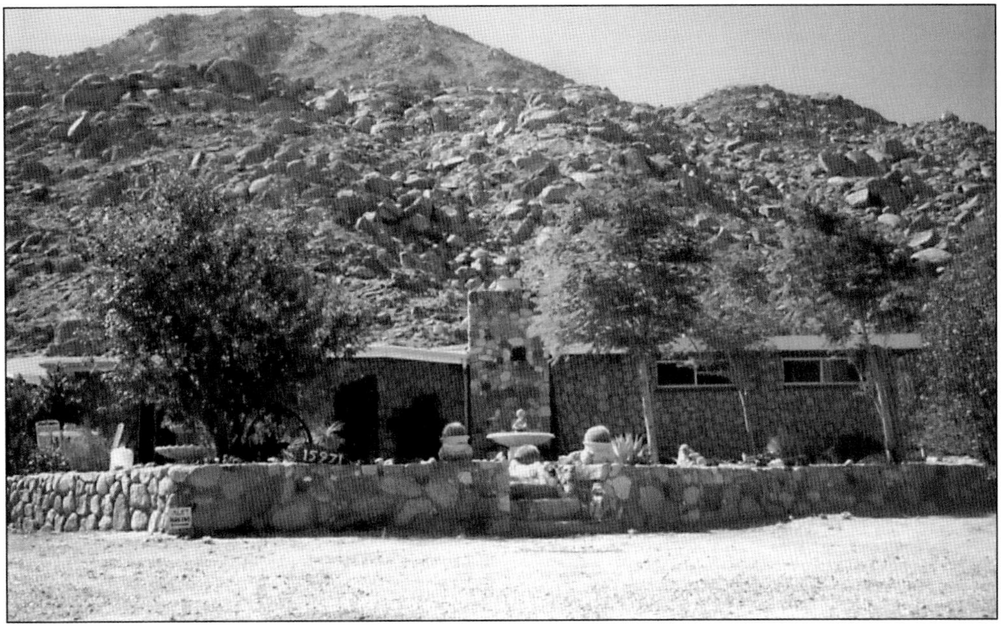

This photograph of a rock house in Snow Creek was taken southwest of Whitewater. Years ago, numerous houses in this area were made of rocks, chiefly because, as can be seen by the rocks on the hill behind, they were an available resource. The more desirable, rounded stones were, however, obtained in river bottoms. (Courtesy of the author.)

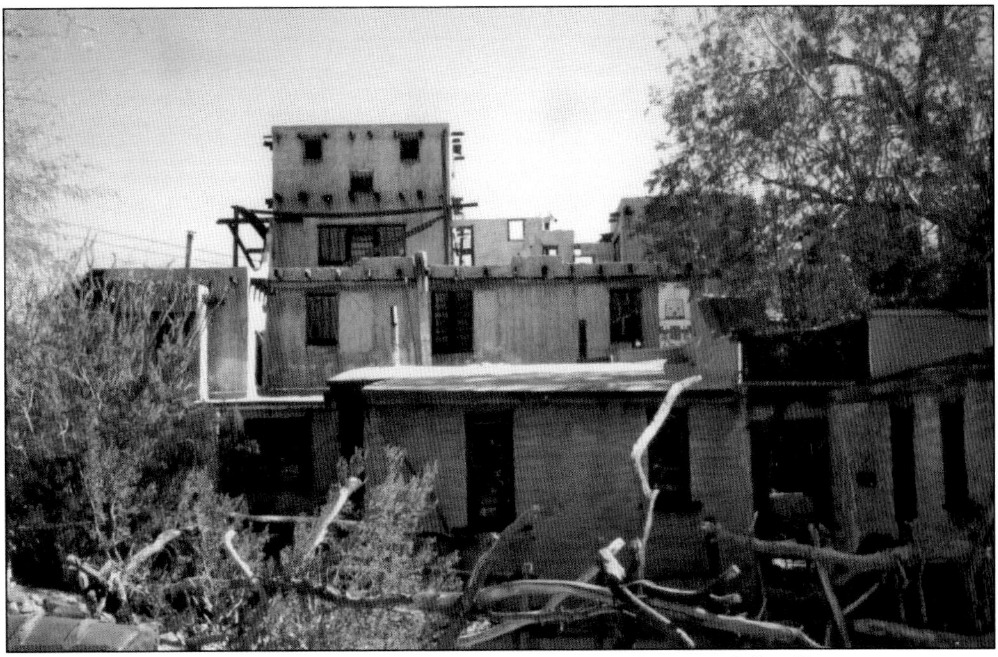

Cabot's Pueblo Museum is east of Whitewater in Desert Hot Springs. Cabot Yerxa was a pioneer in this area and left behind a legacy—his pueblo, a monument to the Native American people he so admired. It is a 35-room, Hopi-style pueblo that contains 150 windows, 65 doorways, and covers an expanse of 5,000 square feet. Cabot spent 23 years of continuous building before he died, leaving it unfinished. (Courtesy of Cabot's Pueblo Museum.)

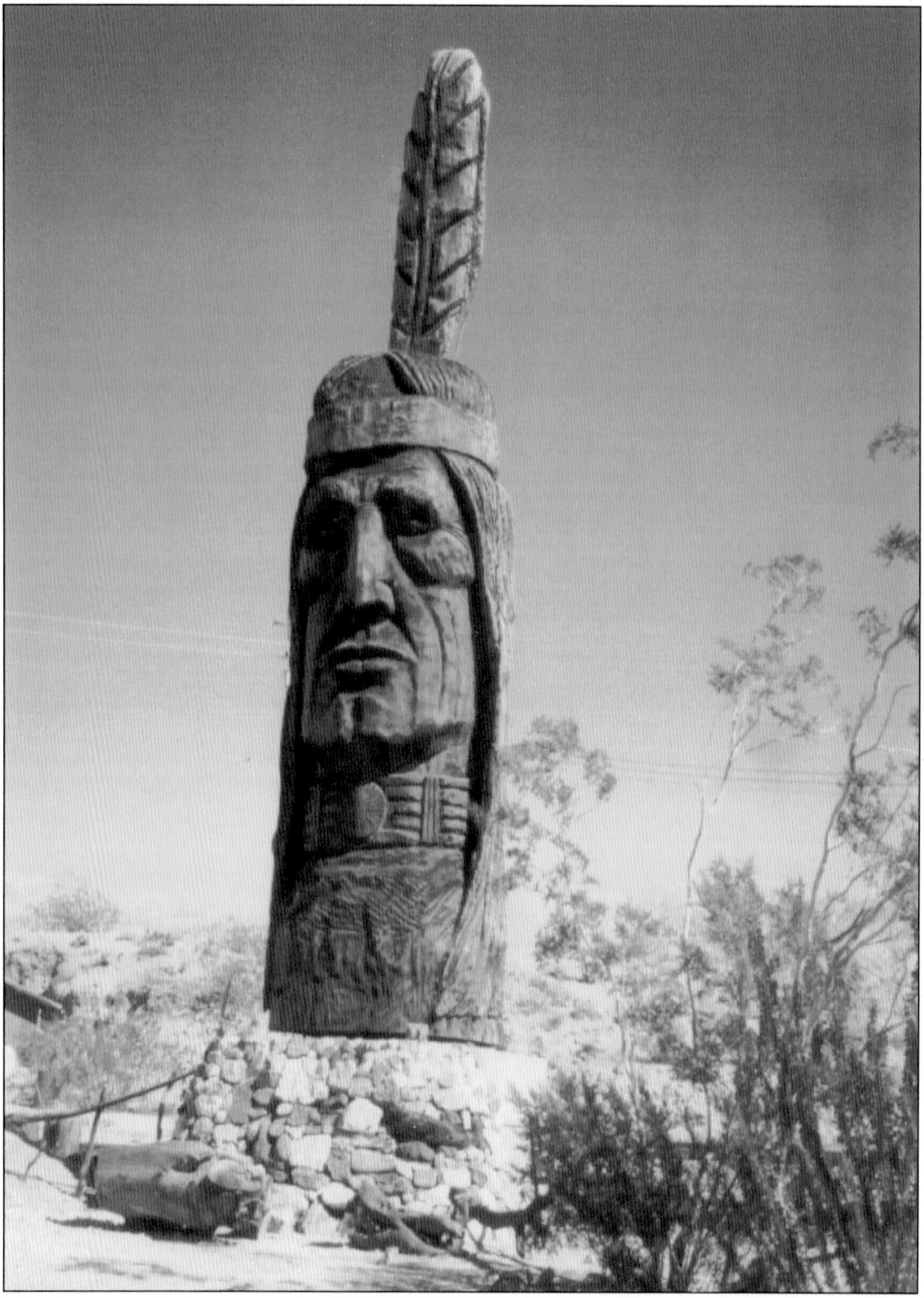

The Cabot Pueblo Museum is home to a 43-foot giant redwood Native American monument. Its massive structure stands high beside the museum and beckons visitors to enter. It was an additional gift, an expression of admiration from Cabot to the Native American people he liked so well. The monument was hand-carved by famed sculptor Peter "Wolf" Toth. (Courtesy of Cabot's Pueblo Museum.)

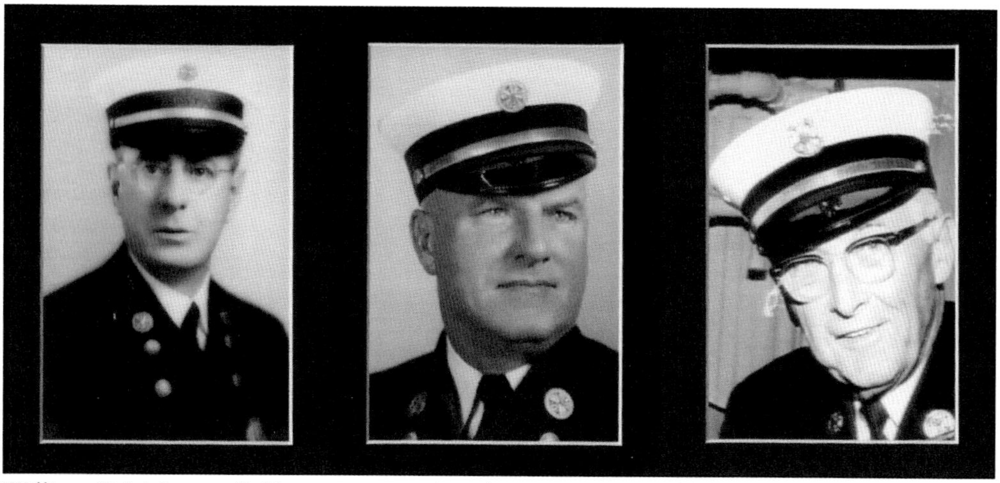

William C. Mahoney (left) was appointed to the new position of full-time chief of the Peabody Fire Department and took command on June 1, 1923, with 22 permanent men and 60 call men serving under him. He served 44 years and died in 1948 while still serving as chief and being a member of many organizations. His funeral was well attended by over 50 chiefs and 800 mourners. Arthur C. Flynn (center) was appointed on November 24, 1948, as the second chief of the Peabody Fire Department. He served 42 years until Edward P. O'Brine (right) was appointed. O'Brine became the third chief of the Peabody Fire Department on September 5, 1962. He served the department a total of 43 years until his retirement and the appointment of Nicholas Gerakaris.

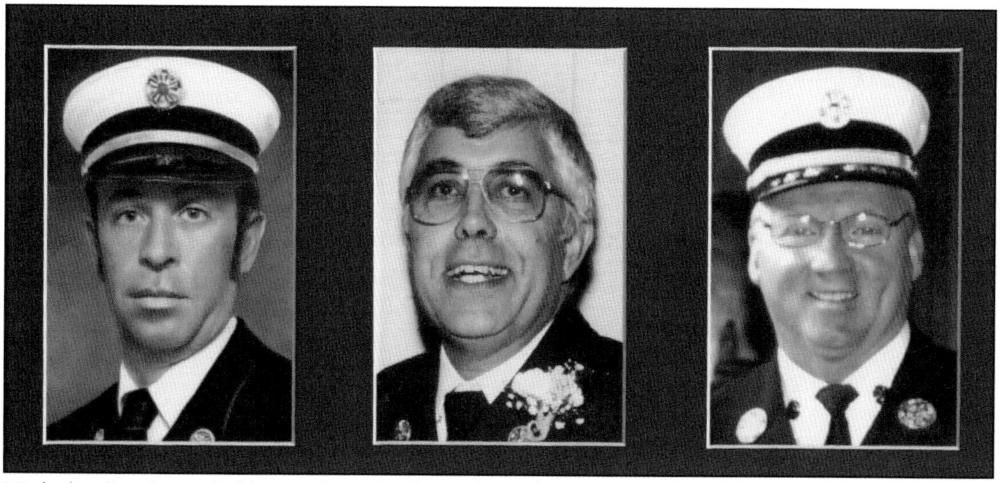

Nicholas Gerakaris (left) rose through the ranks and was promoted to chief on February 29, 1972. While he was chief, he suffered a serious stroke, but he never let it get the best of him, and he continued to serve as chief for 21 years before retiring on January 30, 1993. Joseph A. Mendonca (center) was appointed the fifth chief of the Peabody Fire Department on November 5, 1995. He served the city well for 33 years until his retirement in 2001. Steven Pasdon (right) became the sixth chief of the Peabody Fire Department on November 25, 2001. He is a 20-year veteran of the department.

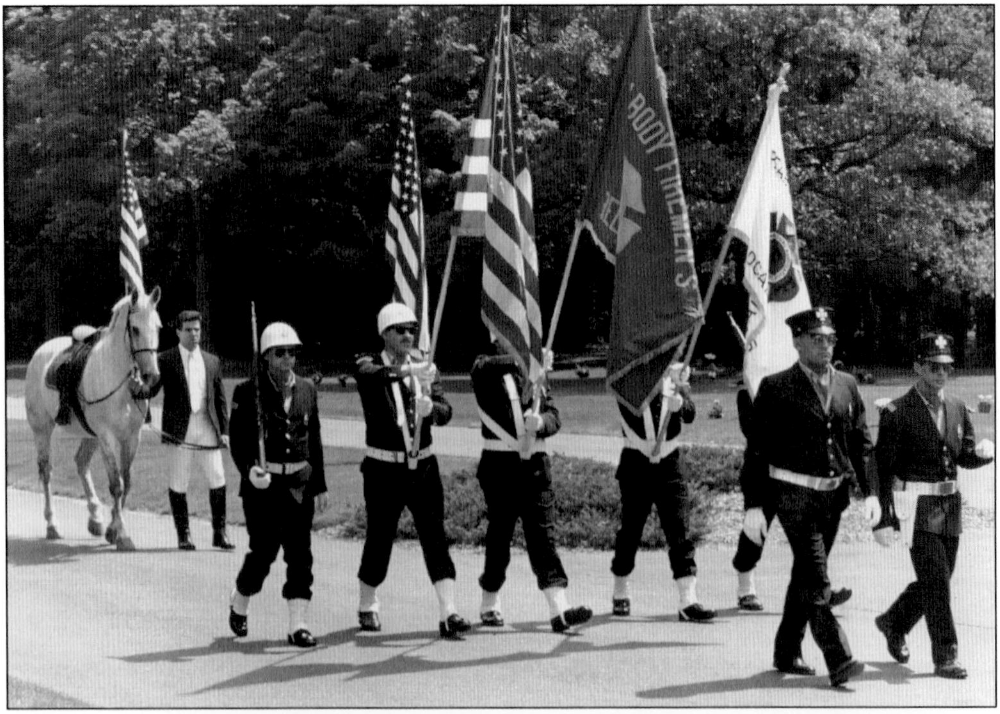

Joe DiFranco (left) and Edward Witcomb lead the Peabody Fire Department honor guard through the Puritan Lawn Cemetery.

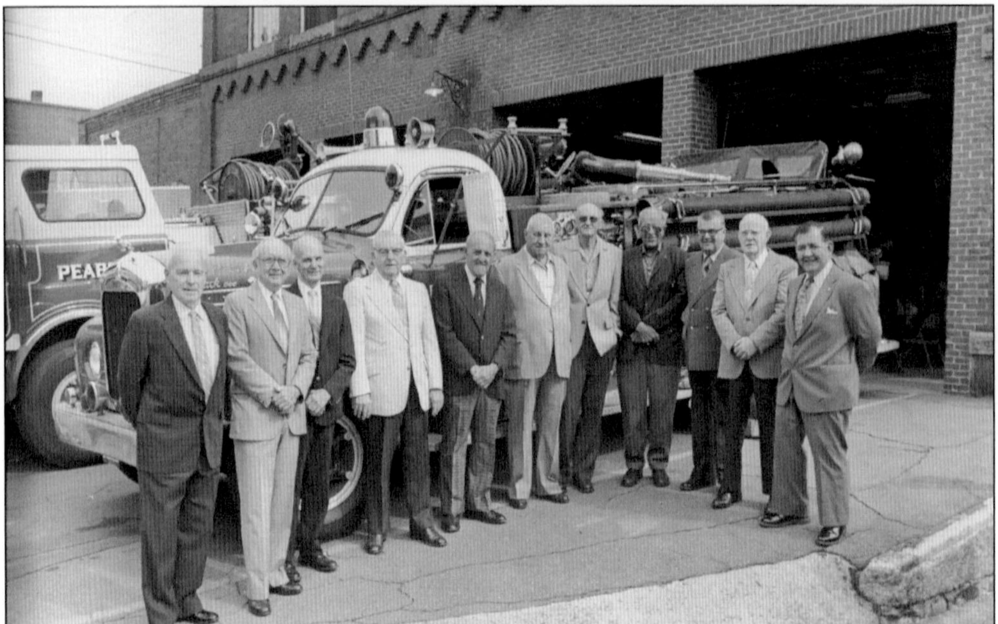

From left to right, Ralph Innis, Jack Quinn, Joseph Horrigan, Lenny Lucey, Bob Dunn, Joseph O'Connor, Jack McCarthy, George Kingsley, Ray Nelson, Charles Kenealy, and Steve Baribo pose in front of Central Station after having breakfast inside Central Station at the annual Firemen's Memorial Sunday observance.

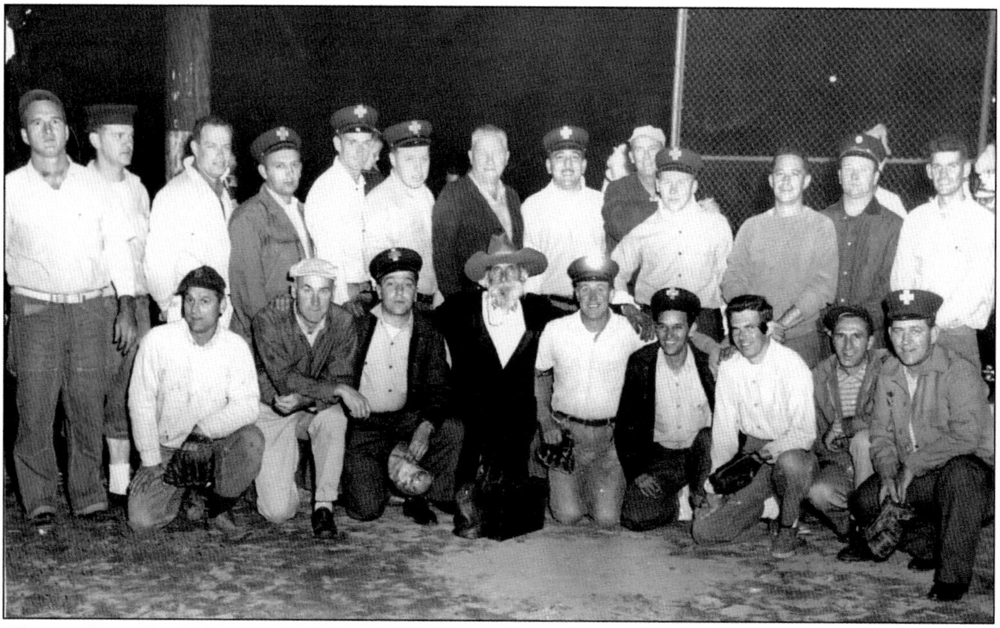

Charity has always been important to the Peabody Fire Department, whether it is for a brother firefighter, muscular dystrophy, the Jimmy Fund, or an annual blood drive. This group photograph was taken in 1963 at one the first baseball games for charity. From left to right are (first row) Maxie Kravchuck, Buddy Leonard, Dan Rosa, Bill Wheeler, unidentified, Arthur Blake, Jospeh Russell, Sal Mistretta and Steve Baribo; (second row) Frank Connors, Bill Lawson, Ralph Putnam, Lenny Marshall, Tom Moran, Dave McDonald, Joseph O'Connor, Tony Koutzoukis, Jack McCarthy, Lou Rochon, Nicholas Gerakaris, Donald McCauley, and Sid Durkee.

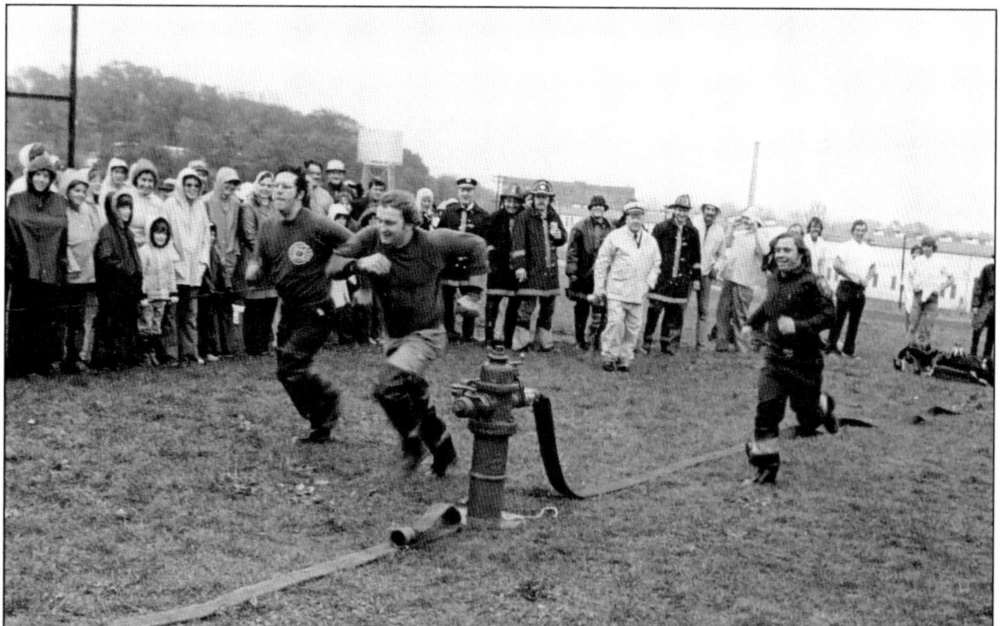

From left to right, firefighters John D'Angelo, Kevin Lynch, and Frank D'Amico Jr. race to the finish line in a competition at the Higgins Middle School field on King Street Extension.

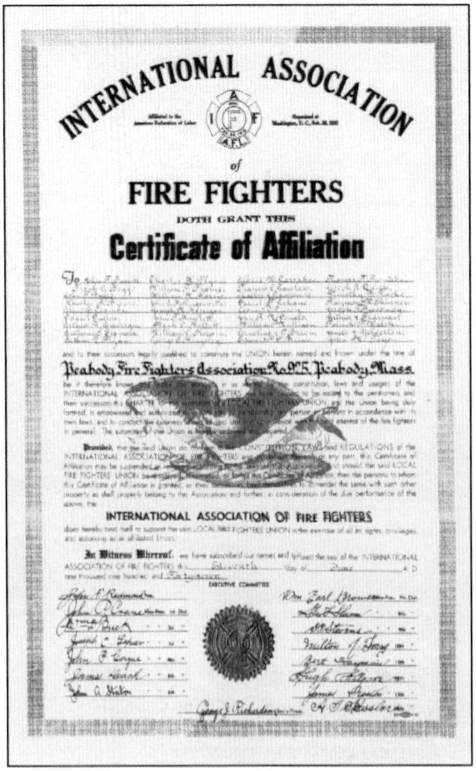

INTERNATIONAL ASSOCIATION
of
FIRE FIGHTERS
DOTH GRANT THIS
Certificate of Affiliation

Peabody Fire Fighters Association No. 925, Peabody, Mass.

INTERNATIONAL ASSOCIATION OF FIRE FIGHTERS

On June 11, 1947, the Peabody Local No. 925 of the International Association of Fire Fighters was organized with John Toye as its first president. The firefighters' union has always advocated for the safety and concerns of its membership.

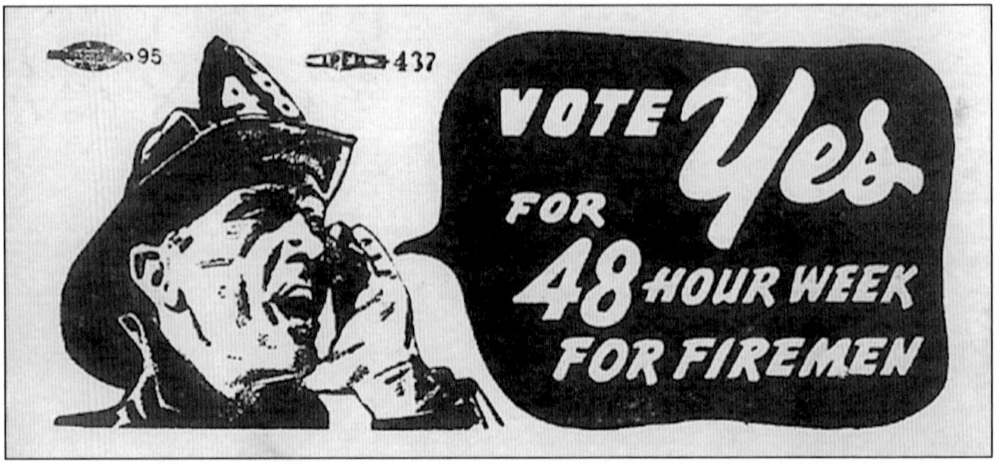

An example of the important role of the firefighters' union was that for many years the firemen worked extremely long hours putting the citizens and themselves in danger. On February 15, 1920, the Two Platoon system went into effect, providing 16 permanent drivers split into two shifts—a day and a night. They worked 14 hours a day for five days straight and then they were entitled to a day off. The Two Platoon system went on for many years until the firemen fought for and won, by voter referendum in 1947, the new 48-hour workweek starting in February 1948. This lasted until a new shift was instituted in the 1970s, consisting of two 10-hour days followed by two 14-hour nights with four days off.

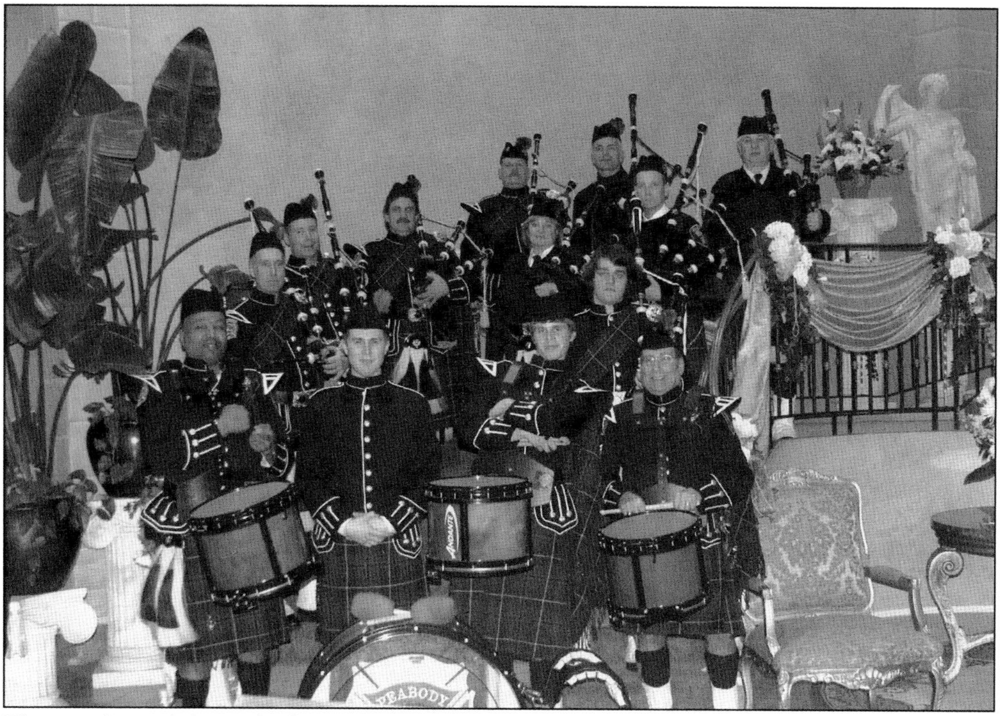

The members of the Peabody Fire Pipes and Drums pose for a picture at Spinelli's restaurant in January 2007 for the annual retirement party. George Nelson, Steve Smyrnios, and John Kerwin were honored that year.

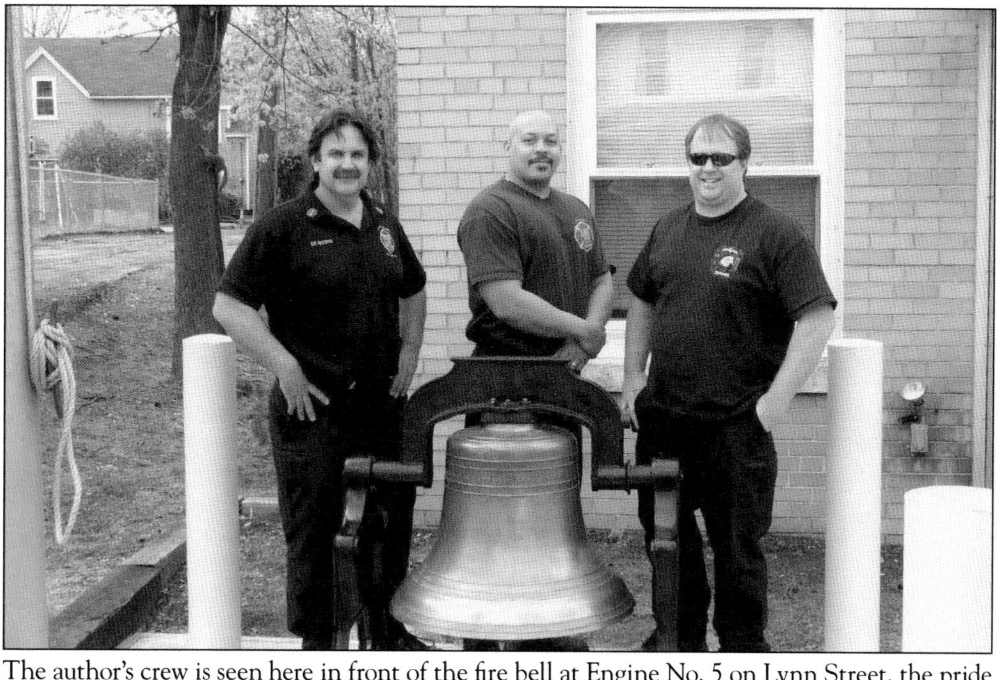

The author's crew is seen here in front of the fire bell at Engine No. 5 on Lynn Street, the pride of the south side. From left to right are Lt. Ted Quinn, firefighter Jimmy Kimber, and firefighter Billy Jacavanco.